JIESUAN DUI CAOYU SHENGZHANG XINGNENG JI
CHANGDAO JIEGOU YINGXIANG DE YANJIU

芥酸对草鱼生长性能及肠道结构影响的研究

甘 雷 著

中国纺织出版社有限公司

图书在版编目（CIP）数据

芥酸对草鱼生长性能及肠道结构影响的研究 / 甘雷著 . --北京：中国纺织出版社有限公司，2025.5.
ISBN 978-7-5229-2618-6

Ⅰ．S965.112

中国国家版本馆 CIP 数据核字第 2025KE1058 号

责任编辑：闫　婷　　责任校对：高　涵　　责任印制：王艳丽

中国纺织出版社有限公司出版发行
地址：北京市朝阳区百子湾东里 A407 号楼　邮政编码：100124
销售电话：010—67004422　传真：010—87155801
http://www.c-textilep.com
中国纺织出版社天猫旗舰店
官方微博 http://weibo.com/2119887771
三河市宏盛印务有限公司印刷　各地新华书店经销
2025 年 5 月第 1 版第 1 次印刷
开本：710×1000　1/16　印张：8.75
字数：152 千字　定价：98.00 元

凡购本书，如有缺页、倒页、脱页，由本社图书营销中心调换

前　言

随着全球水产养殖业的蓬勃发展，水产品已成为许多国家和地区重要的动物蛋白来源。然而，在追求产量提升和效益优化的过程中，水产养殖业仍面临诸多挑战。其中，鱼粉价格的不稳定性导致水产饲料成本波动幅度大是众多难题之一。为应对这一难题，植物蛋白替代鱼粉成为一种行之有效的解决方案。然而，这一替代方案又带来了新的问题，如抗营养因子的负面影响。抗营养因子是指能抑制或干扰动物正常生理功能的物质，它们通过多种途径对鱼类健康产生负面影响，包括阻碍营养物质的消化吸收、干扰代谢过程以及直接破坏肠道结构，最终导致鱼类生长受阻和健康状况下降。常见抗营养因子包括芥酸、植酸、单宁和蛋白酶抑制剂等，这些成分在菜籽粕、豆粕和棉粕等常见植物性蛋白源中所占比例较高。水产饲料中大量使用植物蛋白替代鱼粉，必然导致鱼类遭受抗营养因子的侵扰。随着水产养殖业的现代化进程，人们除了关注鱼类的生长速度和产量，鱼类的健康状况也逐渐成为关注重点。因此，开展抗营养因子对鱼类肠道影响的相关研究，不仅能为降低水产养殖成本提供理论依据，还能为水产品的市场竞争力提升打下坚实基础，已然成为当前水产饲料配方研究的重点课题。

草鱼作为一种重要的水产养殖品种，广泛分布于全球。然而，人们在草鱼的健康管理和养殖效益方面仍面临诸多挑战，例如，植物性蛋白源饲料中存在大量抗营养因子。其中，芥酸作为一种常见的植物来源脂肪酸，广泛存在于富含菜籽粕的饲料中，因其潜在的毒性而长期受到关注。研究表明，芥酸可能通过影响草鱼肠道细胞间结构完整性及内质网应激通路，成为菜籽粕替代鱼粉过程中导致草鱼健康问题的关键因素之一。肠道是草鱼消化吸收和免疫防御的核心器官，其健康状况直接关系到鱼体的生长、免疫力以及生产效益。因此，研究芥酸对草鱼肠道结构及内质网应激信号通路的影响，不仅有助于揭示芥酸对草鱼健康的潜在危害，也为优化草鱼饲料配方、减少饲料中抗营养因子的负面影响提供理论依据。

本书旨在为相关专业高校师生及研究人员提供理论支持与实践指导，同时为水产养殖公司和技术人员提供参考。尽管编者已尽力确保内容的科学性与实用性，但由于研究水平有限，书中可能仍存在疏漏和不足之处。诚恳希望同行专家和读者提出宝贵意见，以进一步完善和提升本书的质量，推动相关领域的研究与实践发展。

著者

2025 年 2 月

目　　录

第一章　绪论 ·· 1
  1.1　引言 ··· 1
  1.2　概论 ··· 3
    1.2.1　菜籽粕中抗营养因子对鱼类影响的研究进展 ······················· 3
    1.2.2　芥酸简介 ·· 5
    1.2.3　芥酸研究现状 ·· 6
    1.2.4　芥酸对动物肠道细胞结构完整性的影响 ···························· 8
    1.2.5　芥酸对动物肠道细胞间结构完整性的影响 ·························· 9
    1.2.6　芥酸对动物内质网应激的影响 ··································· 11
    1.2.7　内质网应激及其介导的信号途径与 RhoA 间的关系 ················ 12
  1.3　存在问题 ··· 13
  1.4　研究内容 ··· 13
  1.5　研究目的和意义 ·· 14
  1.6　技术路线 ··· 14

第二章　芥酸对生长中期草鱼生产性能和肠道结构完整性的影响 ············· 15
  2.1　材料与方法 ·· 16
    2.1.1　试验设计 ··· 16
    2.1.2　饲料配方 ··· 16
    2.1.3　试验条件与饲养管理 ·· 17
    2.1.4　肠道样品采集 ··· 18
    2.1.5　消化试验 ··· 18
    2.1.6　指标测定 ··· 18
    2.1.7　统计分析 ··· 26
  2.2　试验结果 ·· 26
    2.2.1　芥酸对生长中期草鱼生产性能的影响 ···························· 26

2.2.2 芥酸对生长中期草鱼营养物质表观消化率的影响 …… 28
2.2.3 芥酸对生长中期草鱼肠道表观症状和组织结构的影响 …… 28
2.2.4 芥酸对生长中期草鱼血清二胺氧化酶活性和
　　　D-乳酸含量的影响 …… 28
2.2.5 芥酸影响生长中期草鱼肠道结构完整性的可能机制 …… 30
2.3 讨论 …… 41
2.3.1 芥酸降低生长中期草鱼的生产性能 …… 41
2.3.2 芥酸降低生长中期草鱼营养物质表观消化率 …… 41
2.3.3 芥酸破坏生长中期草鱼肠道结构完整性 …… 42
2.3.4 芥酸增加生长中期草鱼血清中二胺氧化酶活性和
　　　D-乳酸含量 …… 43
2.3.5 芥酸对生长中期草鱼肠道细胞结构完整性的影响 …… 43
2.3.6 芥酸对生长中期草鱼肠道细胞间结构完整性的影响 …… 48
2.3.7 生长中期草鱼饲料中芥酸的控制剂量 …… 52
2.4 小结 …… 53

第三章　芥酸激活 RhoA 信号途径 …… 54

3.1 材料与方法 …… 55
3.1.1 试验设计 …… 55
3.1.2 试验材料 …… 55
3.1.3 芥酸配制 …… 56
3.1.4 RhoA 抑制剂配制 …… 56
3.1.5 细胞处理 …… 57
3.1.6 观测指标 …… 57
3.1.7 统计分析 …… 58
3.2 试验结果 …… 58
3.3 讨论 …… 61
3.4 小结 …… 62

第四章　芥酸对内质网应激的影响 …… 63

4.1 材料与方法 …… 64
4.1.1 芥酸对生长中期草鱼肠道内质网应激的影响 …… 64

  4.1.2 内质网应激在芥酸破坏草鱼肠道细胞间结构完整性中的作用 …… 65
4.2 试验结果 ……………………………………………………………………… 67
  4.2.1 芥酸对生长中期草鱼肠道细胞超微结构的影响 ………………… 67
  4.2.2 芥酸对生长中期草鱼肠道内质网应激相关基因表达的影响 …… 67
  4.2.3 芥酸对生长中期草鱼肠道内质网应激及非折叠蛋白反应关键蛋白
     表达的影响 ………………………………………………………… 69
  4.2.4 芥酸对草鱼肠细胞超微结构的影响 ……………………………… 70
  4.2.5 芥酸对草鱼肠细胞内质网应激相关基因 mRNA 水平的影响 …… 71
  4.2.6 芥酸对草鱼肠细胞内质网应激相关蛋白表达的影响 …………… 71
4.3 讨论 …………………………………………………………………………… 74
  4.3.1 芥酸对生长中期草鱼肠道内质网应激的影响 …………………… 74
  4.3.2 内质网应激在芥酸破坏草鱼肠道细胞间结构完整性中的作用 … 76
4.4 小结 …………………………………………………………………………… 77

第五章　内质网应激介导的非折叠蛋白反应 ……………………………………… 78
5.1 材料与方法 …………………………………………………………………… 79
  5.1.1 PERK/eIF2α 信号途径在芥酸影响草鱼肠细胞间结构
     完整性中的作用 …………………………………………………… 79
  5.1.2 IRE1/XBP1 信号途径在芥酸破坏草鱼肠细胞间结构
     完整性中的作用 …………………………………………………… 81
  5.1.3 ATF6 信号途径在芥酸破坏草鱼肠细胞间结构完整性中的作用 … 82
5.2 试验结果 ……………………………………………………………………… 84
  5.2.1 PERK/eIF2α 信号途径在芥酸破坏草鱼肠细胞间结构完整性
     中的作用 …………………………………………………………… 84
  5.2.2 IRE1/XBP1 信号途径在芥酸破坏草鱼肠细胞间结构完整性
     中的作用 …………………………………………………………… 87
  5.2.3 ATF6 信号途径在芥酸破坏草鱼肠细胞间结构完整性中的作用 … 90
5.3 讨论 …………………………………………………………………………… 92
  5.3.1 PERK/eIF2α 信号途径在芥酸破坏草鱼肠细胞间结构完整性
     中的作用 …………………………………………………………… 92
  5.3.2 IRE1/XBP1 信号途径在芥酸破坏草鱼肠细胞间结构完整性
     中的作用 …………………………………………………………… 93

5.3.3 ATF6 信号途径在芥酸破坏草鱼肠细胞间结构完整性中的作用 … 94
5.4 小结 … 95

第六章 结论与展望 … 96
6.1 结论 … 96
6.2 创新点 … 97
6.3 有待进一步研究的问题 … 97

参考文献 … 98

第一章 绪论

1.1 引言

菜籽粕是鱼类饲料中重要的植物蛋白源之一。研究表明，大量添加菜籽粕到水产饲料中会显著降低鲫鱼（*Carassius auratus gibelio* ♀ × *Cyprinus carpio* ♂）、真鲷（*Pagrus major*）、苏里拟鲿（*Pseudobagrus ussuriensis*）、吉富罗非鱼（*Oreochromis niloticus*）、军曹鱼（*Rachycentron canadum*）和草鱼等鱼类的生产性能，并改变异育银鲫（*Carassiusauratus gibelio*）和鲫鱼等鱼类的肠道结构。菜籽粕对鱼类的这些负面影响，可能与其所含的大量抗营养因子有关，芥酸是菜籽粕中的主要抗营养因之一。已有研究发现，芥酸会降低羊羔和小鸡等陆生动物的生产性能，且会增加人多形核白细胞中氧自由基（ROS）的含量，而过量的 ROS 会导致大鼠肠道组织损伤。动物生产性能的下降与肠道健康受损有关，而肠道健康受损又与肠道结构完整性的破坏密切相关。目前，芥酸对水生动物肠道组织结构的影响及机制尚未得到系统深入的研究，因此有必要开展相关实验进行深入研究。

动物肠道细胞间结构完整性及细胞结构的破坏，是导致肠道物理屏障功能受损的关键因素，该过程涉及紧密连接、黏附连接、细胞氧化损伤和凋亡等多个方面。前期研究表明，大豆球蛋白、棉酚和单宁等抗营养因子都会降低鱼类肠道紧密连接蛋白 *ZO-1* 和 *occludin* 的基因表达，并下调抗氧化相关酶基因的 mRNA 水平，同时诱导凋亡指标 DNA 片段化的发生。类似的研究还发现，植酸能下调人结肠癌细胞 Caco-2 中紧密连接蛋白 occludin 和 ZO-1 的蛋白表达。Duan 等（2019）的研究表明，β-伴大豆球蛋白会导致幼草鱼肠道 DNA 片段化的发生，并增加凋亡相关酶 caspase-3、caspase-8 和 caspase-9 的活性。然而，芥酸对动物肠道结构的具体影响及其相关机制尚未有研究报道。已有研究发现，芥酸能降低仔猪心脏中肉碱的含量，而肉碱含量的降低会下调肠道紧密连接蛋白 *ZO-1* 的基因表达。此外，芥酸还会增加大鼠心脏中的花生四烯酸含量，而花生四烯酸的增加会导致人乳腺上皮细胞黏附连接蛋白 E-cadherin 的表达下降，并引发多形核嗜中性粒细胞的凋亡。芥酸还可增加叙利亚金黄地鼠血浆中的葡萄糖含量，而高水平的葡萄糖会激活大鼠肾小球系膜细胞 RhoA 的活性。此外，芥酸会导致大鼠心脏脂肪的蓄积，而高脂肪水平会降低草鱼肠道 CuZnSOD 活性。鼠上的研究发现：芥酸在肝脏中会转化成硬脂酸，而硬脂酸会促进胰腺中胰岛素的分泌；饲喂高水平的胰岛素会引起杂合秋田小鼠肾近端肾小管细胞中 $Nrf2$ 基因表达下调。同时研究还发现：芥酸在大鼠肝脏中能被代谢成油酸，而高水平的油酸会激活人原代滋养细胞中 p38MAPK 和 JNK 信号通路。然而芥酸引起氧化损伤和凋亡作用的信号分子机制未见研究报道，开展芥酸对水生动物肠道结构破坏的机制研究具有重要意义。

据报道，内质网应激的发生会导致动物肠道物理屏障功能受损。内质网应激是动物应对外界刺激的一种适应性机制，通常通过非折叠蛋白反应清除内质网腔中错误折叠或未折叠的蛋白质，以减轻外界应激。然而，若应激未能及时缓解，则会对动物产生负面影响，包括脂肪沉积、细胞凋亡以及肠道屏障功能受损等。已有研究表明，棉酚和单宁会分别引起虹鳟（*Scophthalmus maximus* L.）原代肌肉细胞和人前列腺癌细胞内质网应激。然而，芥酸是否能引起动物内质网应激反应，目前仍不清楚。已有研究表明，芥酸会增加大鼠心脏中游离脂肪酸的水平，而高水平的游离脂肪酸会引发胰腺 β-细胞的内质网应激。因此，研究芥酸是否能够引起内质网应激，并探索其可能的机制，具有重要意义。据报道，内质网应激会降低大鼠小肠隐窝上皮细胞中 E-cadherin 的表达、大鼠血脊髓屏障中紧密连接蛋白 occludin 和黏附连接蛋白 β-catenin 的表达以及人气管上皮细胞中紧密连接蛋白 ZO-1 和黏附连接蛋白 E-cadherin 的表达。然而，芥酸是否通过内质网应激破坏动物肠道细胞结构完整性及其机制尚未明确，因此亟须进一步研究。

草鱼是我国养殖量最大的一种淡水鱼类。由于鱼粉资源的短缺，菜籽粕等植物蛋白逐渐成为鱼类饲料中的主要蛋白源，但其高效利用面临一定的挑战。因此，本试验以草鱼为模型，探讨芥酸对鱼类的影响及潜在机制，具有重要的科学和应用价值。本研究首先通过生长试验，考察芥酸对生长中期草鱼生产性能、营养物质表观消化率、肠道组织结构、血清二胺氧化酶活性及 D-乳酸含量的影响，初步探讨芥酸对草鱼造成负面影响的可能原因。其次，深入探讨芥酸对生长中期草鱼肠道结构（紧密连接、黏附连接、抗氧化和凋亡）的影响及其可能的调控途径（RhoA、Nrf2 和 p38MAPK 信号途径），以揭示芥酸破坏生长中期草鱼肠道细胞结构完整性的潜在机制；再次，通过细胞试验，阻断 RhoA 信号途径，研究芥酸对草鱼肠细胞间结构完整性（紧密连接和黏附连接）的影响，确定芥酸作用的分子机制；最后，结合细胞研究模型，进一步探究内质网应激及其介导的非折叠蛋白反应（PERK/eIF2α、IRE1/XBP1 和 ATF6 信号途径）在芥酸破坏草鱼肠道细胞间结构完整性（紧密连接、黏附连接和 RhoA 信号分子）中的作用。通过这一系列研究，旨在为芥酸对草鱼肠道细胞结构完整性损害的机制提供新的证据，并为菜籽粕在草鱼饲料中的合理使用提供理论依据。

1.2 概论

1.2.1 菜籽粕中抗营养因子对鱼类影响的研究进展

菜籽粕作为一种常见的植物蛋白源，广泛应用于鱼类饲料中。然而，研究表明，大量添加菜籽粕可能对鱼类的生产性能产生不利影响，尤其是对鲫鱼、真鲷、苏里拟鲮、吉富罗非鱼、军曹鱼和草鱼等鱼类的生产性能有明显抑制作用，并改变异育银鲫和鲫鱼等鱼类的肠道结构。这种负面影响可能与菜籽粕中含有多种抗营养因子密切相关。菜籽粕中的主要抗营养因子包括芥酸、硫代葡萄糖苷、芥子酸、植酸和单宁等。以下是这些抗营养因子对鱼类影响的研究现状总结。

1.2.1.1 芥酸

芥酸（顺-13-二十二碳烯酸）是一种长链单不饱和脂肪酸，广泛存在于菜籽粕中。关于芥酸对鱼类的影响，至今尚未见相关研究报道。已有研究显示，芥酸对陆生动物如羊羔、猪、小鸡和大鼠具有显著的不利影响，表现为抑制生产性能、降低营养物质的表观消化率，并促进脂肪在体内的积累。然而，芥酸对鱼类的影响机制和具体作用尚不清楚，因此亟须开展相关研究，探讨芥酸对鱼类生长、消化吸收及肠道结构的潜在影响。

1.2.1.2 硫代葡萄糖苷

硫代葡萄糖苷是一类能够损害动物甲状腺功能的物质。目前已知种类超过120种，它们具有共同的化学结构，包括一个 $β-d-$硫代葡萄糖基团，一个磺化肟部分和一个由蛋氨酸、色氨酸或苯丙氨酸衍生的可变侧链。在硫代葡萄糖苷对鱼类的研究中，Skugor 等（2016）发现，在大西洋鲑的饲料中加入≥3.6%的硫代葡萄糖苷后，鱼的增重和肥满度明显下降。Burel 等（2001）的研究表明，含有50%菜籽粕的饲料（硫代葡萄糖苷含量为 40 mol/g）能显著降低虹鳟的增重和饲料效率。相比之下，von Danwitz 和 Schulz（2020）的研究表明，不同硫代葡萄糖苷含量（0、0.1 mol/g 和 0.7 mol/g）对大菱鲆的终末体重、特定生长率及采食量仅呈现下降趋势。这表明硫代葡萄糖苷对不同鱼类的影响剂量存在差异，仍需要进一步的研究以明确其影响机制。

1.2.1.3 芥子酸

芥子酸是一类酚类化合物，菜籽粕中游离酚酸的 70% 以上为芥子酸。关于芥子酸对鱼类的研究，目前仅有少量文献报道。von Danwitz 和 Schulz（2020）的研究发现，使用不同浓度的芥子酸（0、9 mg/kg 和 316 mg/kg 饲料）喂养大菱鲆 50 天后，芥子酸对大菱鲆的终末体重、特定生长率和采食量呈现出下降趋势。虽然初步结果表明芥子酸可能会对鱼类的生产性能产生负面影响，但这一结论仍需要通过更多的实验验证。

1.2.1.4 植酸

植酸是植物中主要的磷储存化合物，通常以植酸盐形式存在，其能与钙、镁、锌、铜和铁等阳离子结合，抑制鱼类对这类矿物质的吸收，并与饲料中的蛋白质、氨基酸和淀粉结合，降低鱼类的消化率和营养利用率，从而影响其生长。研究表明，植酸在不同浓度下对鱼类生长的影响具有显著差异。例如，植酸（≥3.2%）会显著降低生长中期草鱼的生产性能和肠道免疫功能；在幼草鱼上，40 g/kg 的植酸显著抑制了其生长并降低消化酶（如蛋白酶、脂肪酶和淀粉酶）的活性，以及 4.7 g/kg 的植酸会降低该鱼的增重、饲料摄入和饲料蛋白质效率，同时还会降低其氨基酸和矿物元素的表观消化率。此外，植酸（≥13.5 g/kg）对牙鲆（*Paralichthys olivaceus*）的增重和饲料摄入量也有明显抑制作用。然而，也有研究表明，植酸对鱼类的影响不显著，例如，von Danwitz 和 Schulz（2020）在其研究中发现，植酸（0 和 10.6 g/kg）对大菱鲆的生长、采食量等指标没有显著影响。Chowdhury 等（2014）用植酸含量为 0 和 1.13% 的饲料饲喂虹鳟 12 周后，植酸对虹鳟终末体重、饲料摄入和饲料效率也没有显著影响。Khan 等（2013）采用体外生理生化试验的方法，研究了植酸对露斯塔野鲮（*Labeo rohita*）、喀拉鱼巴（*Catla catla*）和麦瑞加拉鲮鱼（*Cirrhinus mrigala*）前肠和后肠中酶活性的影响，发现植酸会显著降低这三种鱼类肠道中的蛋白酶和 α-淀粉酶活性。总体来看，植酸对鱼类的负面影响程度依赖于其剂量和鱼类种类，且不同鱼类对植酸的耐受性差异较大。

1.2.1.5 单宁

单宁是多酚类物质，具有多种化学结构，分为水解单宁和缩合单宁。单宁能够与饲料中的营养成分（如蛋白质，维生素和矿物质）结合，影响动物的消化吸收。在草鱼的研究中，缩合单宁（≥30 g/kg）显著降低了生长中期草鱼的生产性能和粗蛋白、粗脂肪、干物质等的表观消化率。此外，Yao 等（2019）研究发现，水解单宁对草鱼增重无显著影响。而 Omnes 等（2017）的研究表明，单宁（≥20 g/kg）会降低狼鲈（*Dicentrarchus labrax* L.）的生产性能。罗非鱼的研究也表明，单宁（≥15 g/kg）会显著降低其增重和饲料摄入量。Talukdar 等（2019）通过体外生理生化试验发现，单宁显著抑制了露斯塔野鲮的消酶活性，尤其是胰蛋白酶、胰凝乳蛋白酶和总蛋白酶的活性。综上所述，单宁，特别是缩合单宁，对鱼类的生产性能和消化吸收酶活性具有明显的负面影响。

总的来说，目前，关于菜籽粕中抗营养因子对鱼类生长和消化吸收影响的研究已取得一定进展，但仍存在一些空白。首先，虽然已有研究揭示了芥酸、硫代葡萄糖苷、芥子酸、植酸和单宁等抗营养因子对动物生长的抑制作用，但对于这些抗营养因子如何影响鱼类的肠道结构和其具体机制尚未深入探讨。特别是芥酸对鱼类的影响机制仍不明确，亟须开展相关研究，以补充这一领域的空白。其次，抗营养因子对不同鱼种的影响程度及剂量的差异，仍然需要通过进一步的实验研究来确认。这些研究将有助于为菜籽粕在鱼类饲料中的合理应用提供理论支持。

1.2.2 芥酸简介

芥酸（Erucic acid，化学式：$C_{22}H_{42}O_2$），俗称顺-13-二十二碳烯酸，其结构式为 $CH_3—(CH_2)_7—CH=CH—(CH_2)_{11}—COOH$，是一种单不饱和脂肪酸，含有一个位于羧基端第 13 和第 14 个碳原子之间的不饱和双键（图 1-1）。根据脂肪酸的命名规则，芥酸可简写为 22∶1n-9。芥酸在常温状态下呈白色固体，分子量为 338.57 g/mol，沸点为 265℃，熔点为 33.8℃。它不溶于水，但极易溶于乙醇等有机溶剂。芥酸主要存在于十字花科（Crucifera）和金莲花科（Tropaeolaceae）植物中。由于芥酸会抑制动物对营养物质的利用，它常被视为植物蛋白源（菜籽粕）中的一种抗营养因子。除此之外，芥酸还具有较强的疏水性和良好的润滑性能，被广泛用作油漆、润滑剂等工业产品的原料。

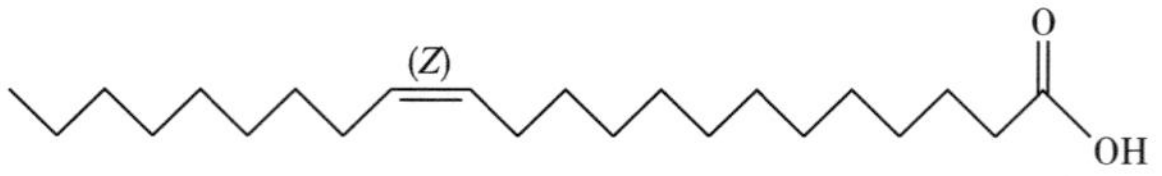

图 1-1 芥酸结构

1.2.3 芥酸研究现状

目前关于芥酸对水生动物的影响尚无研究报道。因此，本书主要总结了芥酸在陆生动物上的研究现状。根据研究方法的不同，芥酸在陆生动物的研究可以分为体内研究和体外研究。本书将围绕芥酸在陆生动物体内试验和体外试验的研究进展进行总结，以期为接下来的实验研究提供一些思路。

1.2.3.1 芥酸在陆生动物体内试验中的研究进展

芥酸在陆生动物体内试验的研究主要涉及3个方面：芥酸对陆生动物生产性能的影响、芥酸对陆生动物营养物质利用的影响，以及芥酸对陆生动物组织器官的影响。以下是这3个方面的研究进展概述。

（1）芥酸对陆生动物生产性能的影响。

动物生产性能是衡量某种营养物质或毒物对动物生理状态影响的重要指标。已有研究表明，芥酸会抑制陆生动物的生产性能。例如，添加 4 g/kg 的芥酸（通过菜籽油添加）会抑制羔羊的生长。类似地，研究发现添加 22.3 g/kg 的芥酸到仔猪饲料中会降低其增重。Renner 等（1979）的研究也表明，5 g/kg 的芥酸会抑制小鸡的生长。Zhang 等（1991）则报道，2.2 g/kg 的芥酸会抑制大鼠的生长。综上所述，芥酸会降低陆生动物的生产性能，但不同物种的最大耐受剂量有所不同。

（2）芥酸对陆生动物营养物质利用的影响。

动物的生长与其对饲料中营养物质的利用能力密切相关。研究表明，芥酸会降低陆生动物对营养物质的表观消化率。例如，芥酸会降低羊羔干物质、粗蛋白和粗脂肪的表观消化率。Sim 等（1985）也发现，芥酸会降低幼鸡总脂肪酸的表观消化率。这些研究表明，芥酸可能通过干扰动物对饲料中营养物质的有效利用，进而影响动物生产性能。

（3）芥酸对陆生动物组织器官的影响。

β-氧化是动物组织中脂肪酸供能的主要途径之一。已有研究发现，芥酸会干扰动物细胞的 β-氧化过程。线粒体是动物细胞进行 β-氧化反应的关键细胞器，尤其在心脏中尤为丰富。大量研究表明，芥酸会对陆生动物的心脏产生负面影响。例如，Astorg等（1981）发现芥酸导致大鼠心脏甘油三酯的蓄积，这一现象在猪和猴身上也得到了类似的验证。此外，Lee和Clandinin（1986）的研究表明，芥酸还会降低大鼠心脏中的ATP利用率。除了心脏，芥酸还可能对其他组织器官产生不良影响。例如，芥酸会降低大鼠肝脏的β-氧化能力，并改变仔猪肝脏细胞内质网和线粒体的形态。Hulan等（1976）发现芥酸会导致大鼠皮肤病变，而在猴体内则可能引起肌肉脂质沉积和心脏肌浆空泡的发生。这些研究结果表明，芥酸对陆生动物的各个组织器官可能产生不利影响，这些负面影响可能是芥酸降低动物生产性能的原因之一。

1.2.3.2　芥酸在陆生动物体外试验中的研究进展

目前，关于芥酸在陆生动物体外试验中的研究主要集中在大鼠的各种细胞（如心肌细胞、肝细胞、脂肪细胞和皮肤成纤维细胞）上，研究内容主要涉及脂肪酸代谢。例如，Pinson等（1974）研究发现芥酸会增加大鼠心肌细胞中神经酸（24：1n-9）和贡多酸（20：1n-9）的含量，这些脂肪酸与芥酸结构相似，其中前者比芥酸多两个碳原子，后者比芥酸少两个碳原子，暗示芥酸可能通过碳链延长和β-氧化转化为其他类型的脂肪酸。Rogers（1977）等的研究也表明，芥酸会改变大鼠心肌细胞膜上的脂肪酸成分。此外，Christensen等（1988）研究发现，芥酸在大鼠皮肤成纤维细胞中可以代谢为油酸（18：1n-9）。这些研究表明，芥酸可能通过碳链延长或缩短，转化为其他类型的脂肪酸，从而改变动物细胞的脂肪酸组成，最终影响组织器官的功能。芥酸还可能直接干扰动物细胞的β-氧化过程，影响细胞的正常功能。Norseth和Christophersen（1978）发现，芥酸抑制了大鼠肝细胞中脂肪酸的β-氧化效率。Christiansen等（1985）的研究也表明，芥酸会对大鼠肝细胞产生脂质毒性效应，导致脂肪酸在细胞内蓄积。脂肪在非脂肪组织中的过度积累可能会引发脂质毒性效应。因此，芥酸可能通过干扰脂肪酸代谢过程，导致动物细胞内脂肪酸的蓄积，从而对动物产生负面影响。总的来说，关于芥酸在陆生动物生长、脂肪酸代谢和组织器官的影响方面，已有大量研究。根据现有的研究成果，我们可以将芥酸的潜在毒性机制归纳为两点：一是芥酸可能通过影响动物组织器官功能，降低动物对营养物质的利用能力，进而降低生产性能；二是芥酸通过干扰细胞的β-氧化过程，导致脂肪酸的蓄积，引发脂质毒性效应（如非脂肪组织中的脂肪沉积），从而抑制动物的生长。然而，关于芥酸对水生动物的影响尚未明确，目前对陆生动物的研究也主要集中在生长、脂肪酸代谢和营养物质利用等方面，芥酸对陆生动物的分子机制仍缺乏深入的研究，尤其是关于其对动物肠道的影响，这一领域仍需进一步探索。

1.2.4 芥酸对动物肠道细胞结构完整性的影响

动物肠道细胞是动物肠道结构的重要组成部分，其完整性受氧化损伤和细胞凋亡等因素的影响。丙二醛（MDA）和蛋白质羰基（PC）分别是脂质和蛋白质氧化的产物，MDA 和 PC 含量的增加通常意味着氧化损伤的加剧。氧化损伤与动物机体抗氧能力的下降密切相关，而抗氧化能力的下降通常表现为抗氧化酶活性减弱和非酶抗氧化物质的含量降低。在鱼类中，主要的抗氧化酶包括超氧化物歧化酶（SOD）、过氧化氢酶（CAT）和谷胱甘肽过氧化物酶（GPX）等，而非酶性物质如谷胱甘肽（GSH）也发挥重要作用。抗氧化酶的活性受相关基因的调控，在斑马鱼中，肠道的抗氧化酶基因主要由核因子 NF-E2 相关因子 2（Nrf2）信号通路调控。DNA 片段化是动物细胞发生凋亡的标志。半胱天冬酶（caspase）在细胞凋亡过程中具有十分重要的作用，其可分为两类，分别是起始 caspases，如 caspase-2、caspase-8 和 caspase-9，以及效应 caspases，如 caspase-3 和 caspase-7。在鱼上，凋亡的发生与死亡受体途径和线粒体途径的激活有关。Bcl-2 蛋白家族（包括抗凋亡因子 Bcl-2 以及促凋亡因子 Bax）调控线粒体途径，而 Fas 受体及其相关因子则涉及死亡受体途径。据报道，p38MAPK 的激活会上调人肺癌细胞中 Bax 的蛋白表达，以及下调小鼠胚胎成纤维（MEFs）中 Bcl-2 的蛋白表达。同时，JNK 的激活会上调人白血病细胞 HL-60 中 FasL 的蛋白表达，以及 MEFs 中 Bax 的蛋白表达。我们实验室前期的研究结果表明：棉酚和单宁会破坏草鱼肠道细胞结构完整性（引发肠道细胞氧化损伤和凋亡）。然而，有关芥酸对动物肠道细胞结构完整性的研究未见报道。Houtsmuller 等（1970）研究发现芥酸会降低大鼠心脏中的磷脂含量。而磷脂含量的减少会导致草鱼肠道中 GSH 含量的减少以及降低抗氧化酶 CuZnSOD 和 CAT 的酶活性。鼠上的研究发现：芥酸在肝脏中会转换成硬脂酸，而硬脂酸会促进胰腺胰岛素的分泌。高水平的胰岛素会下调杂合秋田小鼠肾近端肾小管细胞中 *Nrf2* 的基因表达。芥酸会增加大鼠心脏中花生四烯酸的含量，而高水平的花生四烯酸会导致人多形核嗜中性粒细胞凋亡的发生。芥酸在大鼠小肠中会被代谢成油酸，而高水平的油酸会上调人原代滋养细胞中 p38MAPK 和 c-Jun 氨基末端激酶（JNK）的表达。然而有关芥酸与动物肠道细胞结构完整性的关系及其可能的机制，还有待研究。

1.2.5 芥酸对动物肠道细胞间结构完整性的影响

动物肠道细胞间结构也是动物肠道结构的重要组成部分，其主要由紧密连接和黏附连接组成。其中，紧密连接由40多种蛋白组成，包括跨膜蛋白（如occludin和claudin蛋白）和外周膜蛋白（如ZO蛋白），前者负责细胞与细胞间接触，后者负责细胞与肌动蛋白细胞骨架连接。如图1-2所示，紧密连接蛋白位于细胞间结构中所有连接组件的顶端，紧密连接具有两个功能：①屏障功能，其可调节离子、水和大分子通过细胞间区域；②围栏功能，其可通过限制脂质在膜内的分布来建立和维持细胞极性。而黏附连接蛋白则主要由两大跨膜蛋白家族构成，分别是nectins和cadherins，这两类蛋白的细胞外区域介导细胞与其相邻细胞的黏附，而细胞内区域则与一系列蛋白相互作用。紧密连接蛋白和黏附连接蛋白共同维持动物肠道细胞间结构的完整，从而抵御肠腔中病原菌、毒素和抗原物质带来的危害，同时也调控药物分子、水、营养物质和离子的摄入，从而维持动物肠道稳态。我们实验室前期的研究结果表明：棉酚和单宁会破坏草鱼肠道细胞间结构完整性。然而芥酸是否会影响动物肠道细胞间结构完整性，目前还不清楚。大鼠上的研究表明：芥酸在肠道中会被代谢成油酸，而高水平的油酸会降低心肌细胞中黏附连接蛋白 α-catenin 和肺组织中紧密连接 ZO-1 和 claudin-5 的蛋白表达。在仔猪上的研究结果表明：芥酸会降低心脏肉碱的含量，而肉碱含量的减少会下调肠道紧密连接蛋白 ZO-1 的基因表达。Sato等（1983）研究发现芥酸会增加大鼠心脏花生四烯酸的含量。而花生四烯酸含量的增多会降低人乳腺上皮细胞黏附连接蛋白 E-cadherin 的蛋白表达。由此可见：芥酸可能会破坏动物肠道细胞间结构完整性。然而，该假设仍有待开展实验进行验证。

如图1-3所示，Rho GTPases 是调控动物细胞间紧密连接蛋白和黏附连接蛋白的关键信号分子，其所涉及的调控机制主要有3种：①刺激肌动蛋白在最初的细胞-细胞接触中的聚集，以聚集/稳定基于 E-钙黏着蛋白的突点是必需的；②刺激 NM IIA 活性，进而导致初始顶端连接扩展/线性化以及改变细胞形状，这对于简单上皮的柱状化是必需的；③细胞极性复合体的组装导致紧密连接和底面-顶侧极性的形成。在人上，GTPases 的 Ras 超家族包含150多种蛋白，分为5个家族：Ras、Rho、Rab、Ran 以及 Arf。Rho 家族至少有25个蛋白成员，其中最具特征的是经典的 Rho 蛋白家族，其主要包括三种蛋白：RhoA、Rac 和 Cdc42，它们不仅在调控肌动蛋白—肌球蛋白动力拉伸中起着重要作用，还参与其他的生化过程，如微管动力、基因转录、细胞周期和囊泡运输等。其中 RhoA 主要调控

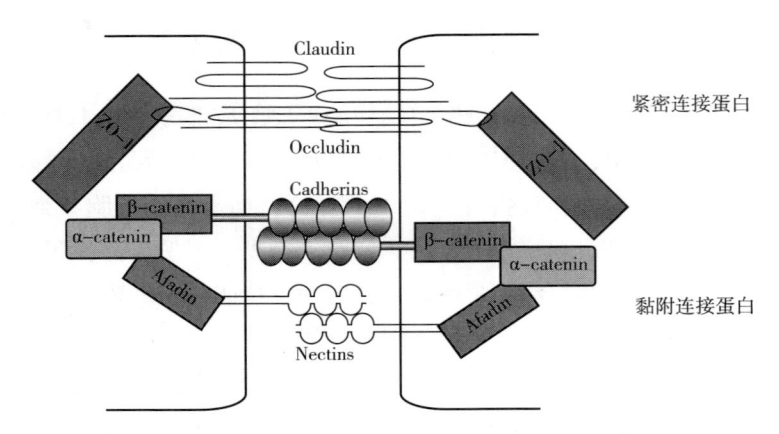

图 1-2　动物肠道细胞间结构

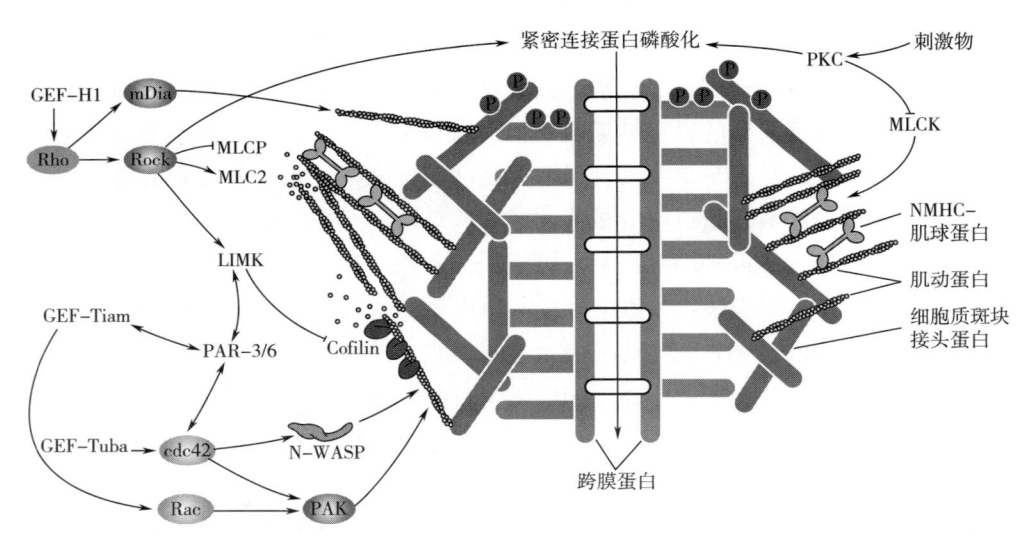

图 1-3　Rho GTPases 与细胞间结构的调节

F-肌动蛋白应力纤维和黏着斑的拉伸，Rac1 控制膜膨出和板层挤压，以及 Cdc42 调节丝状伪足的形成和细胞极性。Nusrat 等（1995）研究发现 Rho GTPase 参与上皮细胞间的紧密连接和结缔组织 F-肌动蛋白的组织调控。Jou 等（1998）进一步确定 Rho GTPase 能参与细胞间紧密连接和黏附连接的调控。为了保持最佳的细胞间连接状态，Rho 蛋白活性需要维持在一个平衡的状态，其活性的过高或过低都会损害上皮屏障功能。由此可见，Rho GTPases 及其效应子的精细调控在上皮细胞间结构完整性的维持中起着关键作用，其中 RhoA 在调控动物细胞间结构完整性又显得尤为重要。研究表明：激活 RhoA 信号途径会抑制小鼠足细胞系中 ZO-1 的蛋白表达和大鼠脑微血管内皮细胞中 ZO-1 和 occludin 的蛋白表达。人结肠腺癌细胞中的研究发现，ZO-1、occludin 和 E-cadherin 蛋白表达的降低与 RhoA 活性的升高有关。然而有关芥酸与 RhoA 的研究未见报道。研究表明：芥酸会加剧人血液中硒缺乏症的发生，而饲料中硒的缺乏会增加小鼠子宫平滑肌 RhoA 信号分子的活性。

根据以上资料，我们发现芥酸可能会通过 RhoA 信号途径破坏动物肠道细胞间结构完整性，然而有关这方面的研究未见报道，因此开展相关研究很有必要。

1.2.6 芥酸对动物内质网应激的影响

内质网是普遍存在于真核细胞中的一种细胞器，其主要参与钙离子的螯合、脂肪合成以及分泌蛋白的翻译、折叠和转运等过程。未正确折叠的蛋白会通过内质网中特定的质量控制机制保留在内质网腔中，然后进行再折叠。如果这一过程还是没有将该蛋白正确折叠，那么该蛋白便会被内质网中的相关适应机制降解掉，以维持细胞内的平衡状态。这种适应性机制，即非折叠蛋白反应（UPR），其维持内质网稳态主要有 4 种方式：①减弱蛋白翻译；②增强内质网蛋白折叠能力；③增加内质网相关的降解能力；④在内质网稳态不能被恢复以及细胞不能克服外界刺激的情况下，其会诱导细胞死亡。这种应激反应主要是通过激活驻留在内质网中的跨膜受体来实现的，这些跨膜受体构成了非折叠蛋白反应中的三条信号途径，分别是：双链 RNA 依赖蛋白激酶样 ER 激酶（PERK）、跨膜蛋白激酶 1（IRE1）和激活作用转录因子 6（ATF6）。在正常动物细胞中，这 3 个内质网跨膜受体可通过各自的腔内区域（IRE1 和 PERK 的氨基末端以及 ATF6 的羧基末端）与内质网分子伴侣—葡萄糖调节蛋白（GRP78）结合，从而使其维持在一个非活性状态。然而，当错误折叠蛋白在内质网腔蓄积后，内质网膜上的这三个跨膜受体便会与 GRP78 分开，进而激活 UPR 信号途径。这三个跨膜受体主要通过信号转导中间体的翻译后修饰（磷酸化或蛋白水解）以及 RNA 修饰（通过 IRE1 核糖核酸内切酶活性）来引发特定的信号级联反应。它们共同参与基因表达的程序调节以恢复内质网的功能，从而维持动物细胞内环境的动态平衡。然而当外界刺激持续时间过长或者刺激强度过大时，这会导致动物细胞内的应激不能被有效解决，UPR 便会对动物细胞产生一系列的负面影响（图 1-4）。据报道，棉酚和单宁都会诱导动物内质网应激。然而芥酸是否会导致动物内质网应激的发生，目前还不清楚。在人类方面的研究表明：芥酸会增加多形核白细胞中活性氧（ROS）的含量，而 ROS 会引发支气管上皮细胞内质网应激。在鼠上的研究表明：芥酸会导致心脏中游离脂肪酸水平的升高，而游离脂肪酸含量的升高会引发胰腺 β-细胞内质网应激。然而，有关芥酸与动物肠道内质网应激的关系，还有待研究。

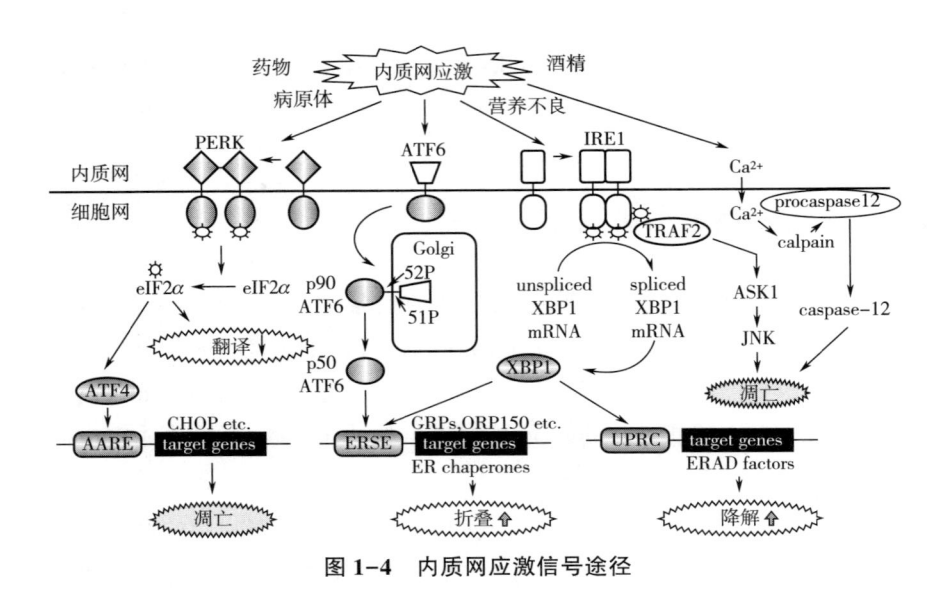

图 1-4 内质网应激信号途径

1.2.7 内质网应激及其介导的信号途径与 RhoA 间的关系

近年来，大量的研究表明：内质网应激会降低大鼠肠道中紧密连接蛋白 ZO-1 和 occludin 的蛋白表达，以及降低人血管上皮细胞中紧密连接蛋白 ZO-1 和黏附连接蛋白 E-cadherin 的蛋白表达，说明内质网应激会破坏动物组织器官细胞间的结构完整性。动物组织器官间的紧密连接和黏附连接受 RhoA 信号分子的调控。因此，我们猜测内质网应激与 RhoA 信号途径间可能存在一定的必然联系。通过查阅的资料也证实了这一设想。例如，Liang 等（2013）研究发现内质网应激会增加小鼠血管平滑肌细胞中 RhoA 的活性。同时，在大鼠 FR3T3 细胞中，内质网应激会激活 RhoA 信号分子。同时，Seo 等（2020）也研究发现内质网应激会上调人表皮角质形成细胞中的 RhoA 活性以及破坏其细胞间的紧密连接。非折叠蛋白反应是内质网应激在动物体内发挥作用的重要方式。非折叠蛋白反应所包含的三条信号途径 PERK/eIF2α、IRE1/XBP1 和 ATF6 都会上调 C/EBP 同源蛋白

（CHOP）的表达。而 CHOP 的激活会上调人骨髓间充质干细胞黏着斑激酶（FAK）的蛋白表达，后者会激活小鼠神经母细胞瘤细胞 RhoA 的活性，说明内质网应激可能会通过其介导的非折叠蛋白应激激活 RhoA 信号途径。近年来，人们也发现非折叠蛋白反应与 RhoA 信号途径间存在密切相关的联系。Xie 等（2019）研究表明敲除 IRE1 会降低人结肠癌细胞 HCT116 中 RhoA 的活性。Seo 等（2020）研究发现人正常表皮角质形成细胞中 RhoA 活性的增加与 PERK 信号途径的激活有关。以上研究结果说明，内质网应激可能会通过激活其介导的非折叠蛋白反应途径激活动物体内的 RhoA 信号分子。然而，内质网应激及其介导的信号途径是否激活 RhoA 信号途径破坏动物肠道细胞间结构完整性，目前还不清楚，开展相关研究很有必要。

1.3 存在问题

（1）关于芥酸对羊羔、仔猪、小鸡和大鼠等陆生动物生产性能的影响有少量研究，而水生动物未见研究报道。

（2）关于菜籽粕影响水生动物肠道结构有少量研究，而菜籽粕中的抗营养因子芥酸对动物肠道结构的影响及作用机制未见研究报道。

（3）关于菜籽粕在草鱼饲料中的控制剂量已有报道，而关于菜籽粕中的抗营养因子芥酸在草鱼饲料中的控制剂量尚不清楚。

1.4 研究内容

本研究结合体内和体外试验，采用组织学、生物化学和分子生物学等研究手段，以草鱼为研究对象，开展如下研究：

（1）考察芥酸对生长中期草鱼生产性能、营养物质表观消化率、肠道组织结构、血清二胺氧化酶活性、血清 D-乳酸含量、肠道紧密连接、肠道黏附连接、肠道氧化损伤和肠道凋亡及相关信号分子的影响，并确定生长中期草鱼饲料中芥酸的控制剂量。

（2）通过体外试验，研究芥酸对草鱼肠道关键信号分子 RhoA 活性、紧密连接和黏附连接相关蛋白的基因和蛋白表达的影响。

（3）通过生长试验和体外试验，研究芥酸对草鱼肠道细胞超微结构和内质网应激相关蛋白的基因及蛋白表达的影响。

（4）通过体外试验，研究内质网应激介导的非折叠蛋白反应途径（PERK/eIF2、IRE1/XBP1和ATF6）在芥酸改变草鱼肠道细胞间结构完整性中所起的作用。

1.5 研究目的和意义

探明芥酸对生长中期草鱼生产性能、肠道结构完整性和内质网应激的作用，并揭示其作用的分子机制，为草鱼对植物蛋白菜籽粕的高效利用提供部分理论依据，具有重要的学术意义；根据生产性能指标确定生长中期草鱼饲料中芥酸的控制剂量，为草鱼饲料配制中菜籽粕的合理使用提供依据，具有重要的生产意义。

1.6 技术路线（图1-5）

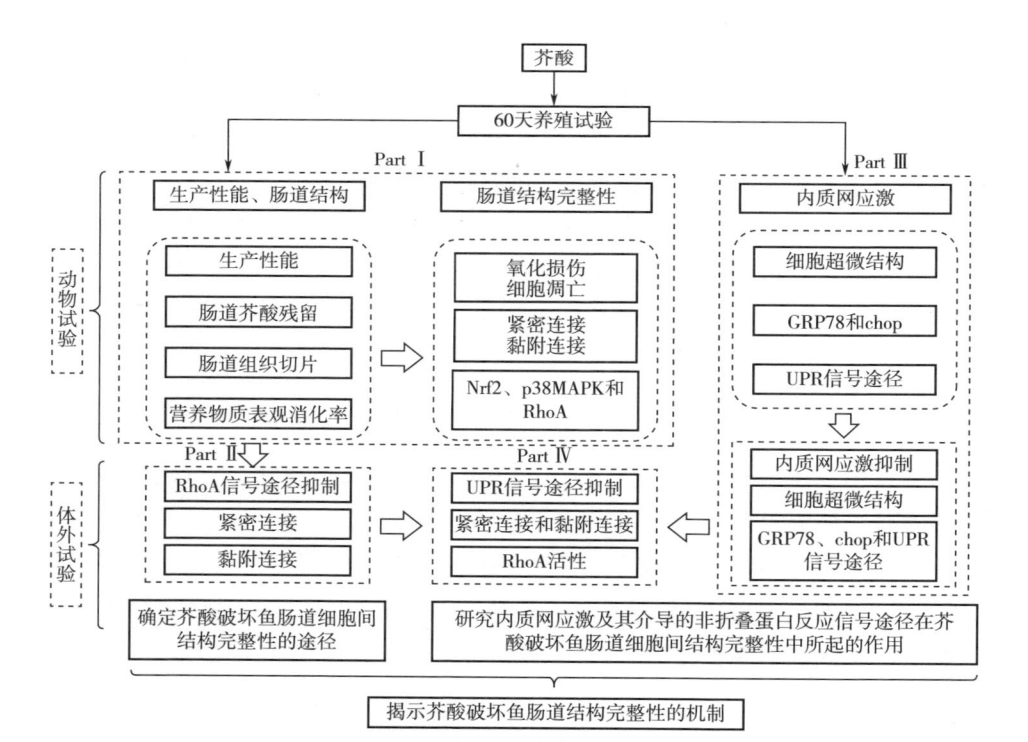

图1-5 整体试验研究思路

第二章 芥酸对生长中期草鱼生产性能和肠道结构完整性的影响

菜籽粕是水产饲料中的一种重要植物蛋白源。研究表明：鱼饲料中大量使用菜籽粕会对鱼类生产性能产生负面影响。这种负面影响可能与菜籽粕中的抗营养因子相关，其中，芥酸作为菜籽粕中的抗营养因子之一，已被证实会抑制羊羔和小鸡的生长，并降低小鸡的营养物质表观消化率。然而，目前关于芥酸的研究大多集中在陆生动物，而水生动物方面的研究尚未见报道。已有研究发现，动物生长与肠道健康密切相关，而动物肠道健康很大程度上依赖于肠道细胞结构的完整。然而，芥酸与动物肠道结构的关系仍未得到充分探讨。已有文献报道，芥酸能够增加人类多形核白细胞中的氧自由基（ROS）含量，而过量的 ROS 会导致大鼠肠道组织损伤。这些研究提示，芥酸可能通过引发氧化损伤，进而破坏动物的肠道结构完整性，但这一假设仍需进一步验证。

在鱼类中，肠道细胞结构的完整性对于维持肠道健康至关重要，细胞间结构和细胞本身的完整性是相互关联的。肠道氧化损伤和细胞凋亡是导致动物肠道细胞间结构完整性破坏的主要因素。斑马鱼的研究表明，肠道的氧化损伤受 Nrf2 信号途径的调控。此外，段绪东（2019）在幼草鱼中的研究表明，β-伴大豆球蛋白能够通过激活 p38MAPK 信号途径诱导草鱼肠细胞的凋亡。细胞间紧密连接和黏附连接是鱼类肠道细胞间结构的重要组成部分，且这些结构主要受 RhoA 信号途径的调控。尽管目前尚无关于芥酸对动物肠道细胞结构完整性以及细胞间结构的研究，但已有研究表明芥酸可能通过影响脂肪代谢、氧化损伤以及调控相关信号途径，间接影响肠道的结构和功能。例如，芥酸会导致大鼠心脏中脂肪的积累，而高水平的脂肪会降低草鱼肠道中 CuZnSOD 的活性。另外，芥酸在大鼠肝脏中代谢为硬脂酸，而硬脂酸会促进胰岛素分泌。高水平的胰岛素能够下调杂合秋田小鼠肾近端肾小管细胞中 *Nrf2* 的基因表达。进一步的研究表明，芥酸会增加大鼠心脏中花生四烯酸的含量，而高水平的花生四烯酸会导致人类多形核嗜中性粒细胞凋亡，并下调人乳腺上皮细胞中黏附连接蛋白 E-cadherin 的蛋白表达。此外，研究还发现，芥酸在大鼠小肠中代谢为油酸，而高水平的油酸能够激活人类原代滋养细胞中的 p38MAPK 和 JNK 信号通路。在仔猪的研究中，芥酸还会减少心脏中的肉碱含量,而肉碱含量的减少会下调肠道中紧密连接蛋白 $ZO-1$ 基因的表达。与此同时,芥酸会增加叙利亚金黄地鼠血浆中葡萄糖的含量,而高水平的葡萄糖会增加大鼠肾小球系膜细胞 RhoA 的活性。这些研究结果说明,芥酸可能会通过调控 Nrf2、p38MAPK(或 JNK)和 RhoA 等信号途径,破坏动物肠道的结构完整性。然而,芥酸对动物肠道结构完整性的影响机制仍需深入研究。

草鱼是我国养殖量最大的一种淡水鱼类。由于鱼粉资源短缺,菜籽粕等植物蛋白成为主要的蛋白来源,但其高效利用面临一定问题。因此,本研究选择草鱼作为实验对象,探讨芥酸对水产动物的影响及可能的机制,对于提高水产饲料中菜籽粕的利用效率,改善鱼类的健康和生产性能具有重要的实际意义。

2.1　材料与方法

2.1.1　试验设计

试验选取 540 尾均重为 129.17 g 左右的生长中期草鱼,随机分为 6 个处理组,每个处理组 3 个重复,每个重复 30 尾鱼。试验设计如表 2-1 所示。

表 2-1　试验设计方案

处理组	1	2	3	4	5	6
每个处理重复数	3	3	3	3	3	3
每个重复试验鱼尾数	30	30	30	30	30	30
芥酸添加水平/%	0.00	0.30	0.60	0.90	1.20	1.50
芥酸实测水平/%	0.00	0.29	0.60	0.88	1.21	1.50

2.1.2　饲料配方

如表 2-2 所示,饲料配方以鱼粉、酪蛋白和明胶为饲料主要蛋白源,以鱼油和大豆油为主要脂肪源。饲料中粗蛋白、粗脂肪以及其他营养成分均能满足生长中期草鱼的生长所需。芥酸(纯度>93%)购于禾大西普化学(四川)有限公司(绵阳,四川,中国)。

第二章 芥酸对生长中期草鱼生产性能和肠道结构完整性的影响

表 2-2 基础饲料配方及营养成分

原料	含量/%	营养成分	含量/%
鱼粉	5.10	粗蛋白[④]	28.73
酪蛋白	23.86	粗脂肪[④]	6.43
明胶	6.50	粗纤维[④]	4.26
DL-蛋氨酸（99%）	0.21	灰分[④]	5.80
L-色氨酸（99.2%）	0.05	n-3[⑤]	1.04
鱼油	2.52	n-6[⑤]	0.96
大豆油	1.81	有效磷[⑥]	0.40
α-淀粉	22.00	总能（cal/g）	4285.19
玉米淀粉	22.39		
微晶纤维素	5.00		
维生素预混料[①]	1.00		
矿物质预混料[②]	2.00		
氯化胆碱（50%）	1.00		
乙氧基喹啉（30%）	0.05		
磷酸二氢钙	1.51		
芥酸预混料[③]	5.00		

[①]每千克维生素预混料含（g·kg^{-1}）：维生素 A 乙酸酯（500000 IU·g^{-1}），0.39；维生素 D3（500000 IU·g^{-1}），0.20；DL-α-生育酚醋酸酯（50%），23.23；维生素 K（22.9%），0.83；维生素 B$_{12}$（1%），0.94；D-生物素（2%），0.75；叶酸（95%），0.17；硝酸硫胺（98%），0.10；维生素 C 醋酸酯（95%），9.77；烟酸（99%），3.44；肌醇酯（98%），28.23；D-泛酸钙（98%），3.85；维生素 B$_2$（80%），0.73；盐酸吡哆醇（98%），0.45。用玉米淀粉补足至 1kg。

[②]每千克矿物质预混料含（g·kg^{-1}）：一水和硫酸锰（31.8% Mn），2.6600；一水合硫酸镁（15.0% Mg），200.0000；一水合硫酸亚铁（30.0% Fe），12.2500；一水合硫酸锌（34.5% Zn），8.2500；五水合硫酸铜（25.0% Cu），0.9560；碘酸钙（3.25% I），1.54；亚硒酸钠（44.7% Se），0.0168。用玉米淀粉补足至 1kg。

[③]通过添加芥酸预混料获得不同芥酸水平处理的饲料，同时各芥酸饲料中通过添加适量的月桂酸以平衡各芥酸处理饲料中的脂肪含量。

[④]粗蛋白、粗脂肪、粗纤维和灰分为测定值。

[⑤]n-3 和 n-6 参考 Zeng 等（2016），以及参考 NRC（2011）进行计算。

[⑥]有效磷参考 Wen 等（2015），以及参考 NRC（2011）进行计算。

2.1.3 试验条件与饲养管理

本实验在四川省成都市大邑县四川畜牧科学研究院实验基地开展。动物的饲养与管理按照本实验室草鱼网箱养殖规范进行。试验用鱼在 1.4 m（长）× 1.4 m（宽）× 1.4 m（高）的网箱中暂养 4 周，以适应饲养环境。暂养期间的水温控制在 26.8±1.6℃，pH 值为 7.0±0.4，溶解氧浓度不低于 6.0 mg/L。暂养期结束后，随机选取 540 尾体重为 129.17±0.19 g 的健康草鱼，并将其均匀分配到 18 个网箱中。每个网箱中悬挂一个直径为 100 cm 的圆形网盘，用于收集剩余饲料。正式饲养试验持续 60 天，采用邵绪远（2019）的方法进行。实验期间，每天分 4 次进行投喂，时间分别为 8：00、11：00、15：00 和 19：00。投喂量根据草鱼的生长速度、采食情况、水温和天气等因素调整，确保每次投喂 30 min 后仍有少量剩料，并将其收集、烘干、称重，用于计算采食量。整个养殖期间，水温、pH 值和溶解氧浓度等养殖环境条件与暂养阶段保持一致。

2.1.4 肠道样品采集

养殖试验结束后，饥饿 12 h，随机从各处理组中选取 9 尾鱼，参考实验室前期的方法，用浓度为 50.0 mg/L 的对氨基苯酸乙酯（纯度>99%，梯希爱，中国）对草鱼进行麻醉处理。麻醉后称重和测量鱼的体长，然后在冰上沿着鱼体侧线将鱼剖开，迅速分离出肠道，剔除肠道上的脂肪等黏附物，并用生理盐水冲洗肠道，拍照、称量肠重和测量肠长。随后，按 Stroband 等（1979）的方法将肠道分为前、中和后肠三个肠段，并立即置于液氮中速冻，然后保存在-80℃条件下备用。另外每个处理组选取 3 尾鱼，采集前、中和后肠，将其置于 4% 多聚甲醛溶液中固定。

2.1.5 消化试验

根据生长试验结果，选取芥酸添加水平为 0.00、0.88% 和 1.50% 的芥酸处理组开展消化试验。在这三个芥酸处理组的饲料中分别按 5 g/kg 的量添加 Cr_2O_3 作为惰性指示剂。消化试验开始前，先让草鱼适应一周，然后选取 270 尾初始体重约为 133.21 g 的健康草鱼随机分配到 9 个试验网箱中，每个网箱 30 尾鱼。正式试验共 10 天，试验期间的养殖条件与生长第二章致。试验结束后，参照 Austreng 等（1978）的方法轻轻挤压鱼体腹部。将收集好的粪便保存在-20℃条件下用于后期的试验分析。

2.1.6 指标测定

2.1.6.1 生产性能指标

养殖试验结束后，记录每个网箱的草鱼尾数并称重，参照 Xu 等（2016）的方法计算草鱼生产性能相关指标：

增重（g 尾$^{-1}$）= 末体重（g 尾$^{-1}$）-初体重（g 尾$^{-1}$）

增重百分比（PWG）（%）= 100×[增重（g 尾$^{-1}$）/初体重（g 尾$^{-1}$）]

特定生长率（SGR）（%）= 100×（ln 末体重-ln 初体重）/试验天数

摄食量（g 尾$^{-1}$）= 投饵量（g 尾$^{-1}$）-残饵量（g 尾$^{-1}$）

饲料效率（FE）（%）= 100×[增重（g 尾$^{-1}$）/摄食量（g 尾$^{-1}$）]

参照 Zhang 等（2013）的方法，根据草鱼体重、体长、肠重和肠长计算草鱼肠道发育指标：

肠体指数（ISI）（%）= 肠道重量（g 尾$^{-1}$）/体重（g 尾$^{-1}$）×100

肠长指数（ILI）（%）= 肠道长度（cm 尾$^{-1}$）/体长（cm 尾$^{-1}$）×100

2.1.6.2 饲料常规营养成分测定

饲料中粗蛋白、粗脂肪和灰分参照 AOAC（2005）的方法进行测定。其中粗蛋白采用凯氏定氮法测定，粗脂肪通过索氏抽提法测定，灰分含量主要是将饲料样品先经过高温炭化（在电炉上进行）后置于马弗炉（550℃）中灼烧至过夜获得，饲料粗纤维含量用酸碱消煮法进行测定，具体操作参考国标法 GB/T 6434—1994。饲料总能采用氧弹量热仪（Parr 6400）进行测定。

2.1.6.3 血清指标检测

饲养试验结束时，参照甘雷（2016）的方法获取草鱼血清。具体操作如下：每个处理选取体重接近平均体重的 6 尾鱼进行尾静脉取血，然后将全血置于常温下凝固 1 h，再在 4℃环境中静置过夜，随后在室温条件下离心（4000 r/min，5 min），吸上清。将血清分装并保存在-80℃条件下备用。根据相应的试剂盒说明书，对草鱼血清中二胺氧化酶（DAO）活性（QS48580，北京奇松，中国）和 D-乳酸含量（QS48279，北京奇松，中国）进行测定。

2.1.6.4 草鱼肠道组织芥酸含量测定

参照 Folch 等（1957）的方法，采用氯仿：甲醇法（2:1）提取各芥酸处理组草鱼肠道中的脂肪，然后根据 Liu 等（2012）的方法，将提取出来的脂肪进行皂化和甲酯化处理。参照 Li 等（2013）方法，采用配备了毛细管柱（30 m×0.25 mm，0.25 μm 膜厚度）和火焰离子检测器的气象色谱仪（GC-2010 plus Shimadzu，Kyoto，Japan）分析草鱼各肠段组织样品中的芥酸含量。以十九烷酸甲酯作为内标，用芥酸甲酯做标准曲线。

2.1.6.5　肠道组织观察

参照 Kokou 等（2017）的方法，将前、中和后肠样品在4%多聚甲醛溶液中固定，然后经石蜡包埋、脱水和切片等处理，用苏木精—伊红（H & E）染色，并在 Nikon TS100 光学显微镜（尼康公司，东京，日本）下观察草鱼肠道组织结构，并拍照。

2.1.6.6　表观消化率测定

各芥酸处理组中饲料和粪便中的干物质、粗脂肪和粗蛋白的测定参照 AOAC（2005）的方法进行。草鱼营养物质表观消化率计算方法参考 Burel 等（2000）：

$ADC_{DM}=100×$ ［1-（饲料中的 Cr_2O_3 含量/粪便中的 Cr_2O_3 含量）］；

$ADC_{营养物质}=100×$ ［1-（饲料中的 Cr_2O_3 含量/粪便中的 Cr_2O_3 含量）×（粪便中的营养物质含量/饲料中的营养物质含量］。

2.1.6.7　抗氧化状态相关指标

称取 0.5 g 的草鱼肠道样品（前、中或后肠）至含有 4.5 mL 预冷的 0.9%生理盐水的 10 mL EP 管中，然后用剪刀将样品剪成食糜状，再超声 2 min 以充分破碎组织，将其变成组织匀浆。随后将组织匀浆液离心（6000 r/min，4℃，20 min），收集上清液备用。

（1）氧化损伤指标。

活性氧（ROS）含量：参考 Jiang 等（2015）的方法。

丙二醛（MDA）含量：参考 Kucukbay 等（2006）的方法。

蛋白羰基（PC）含量：参考 Castex 等（2009）的方法。

（2）酶类和非酶类抗氧化指标。

ASA 和 AHR 活力：参考 Zhang 等（2013）的方法。

CuZnSOD 和 MnSOD 活性：参考 Lu 等（2015）的方法。

CAT 活性：参考 Parrilla-Taylor 等（2013）的方法。

GPx 活性：参考 Trenzado 等（2006）的方法。

GST 活性：参考 Habig 等（1974）的方法。

GR 活性：参考 Carlberg 和 Mannervik（1975）的方法。

GSH 含量：参考 Ritola 等（2002）的方法。

2.1.6.8　DNA 片段化

草鱼肠道 DNA 片段化检测主要参考 Kawakami 等（2008）的方法进行。按每个泳道 1000 ng 的量取适量各芥酸处理组样品的 DNA，经 2%琼脂糖凝胶电泳进行分离（1.5 h，40 V）。然后用凝胶成像系统（Syngene，Frederick，MD，USA）进行拍照。

2.1.6.9　基因表达

（1）试剂和药品。

琼脂糖（sigma，美国）、DNA marker 2000（宝生物，日本）、DEPC 水（上海生工，中国）、RNAiso plus 试剂盒（宝生物，日本）、反转录 PrimeScript™ RT 试剂盒（宝生物，日本）、实时荧光定量 PCR 试剂盒（宝生物，日本）、无水乙醇（上海生工，中国）、氯仿（上海生工，中国）和异丙醇（上海生工，中国）等。

（2）仪器设备。

SCIENTZ-48 高通量组织研磨仪（宁波新芝生物科技股份有限公司，中国）、冷冻离心机（Eppendorf，德国）、移液枪（Eppendorf，德国）、电泳仪（北京六一仪器厂，中国）、电泳槽（北京六一仪器厂，中国）、核酸蛋白检测仪（Thermo，美国）、CFX96 实时荧光定量 PCR 仪（BIO-RAD，美国）、凝胶成像分析仪（BIO-RAD，美国）、普通 PCR 扩增仪（BIO-RAD，美国）、TX223L 电子天平（SHIMADZU，日本）和超净工作台 SB-2（上海净化设备厂，中国）等。

（3）试剂配制及准备。

（a）75%乙醇：取 150 mL 无水乙醇于 50 mL DEPC 水中，混匀，置于冰上预冷备用。

（b）选用优质枪头和 EP 管，并经高温灭菌处理，然后置于烘箱中烘干，备用。

（c）将氯仿和异丙醇提前置于冰上预冷，备用。

（4）试验方法。

（a）引物设计。

荧光定量引物是根据本实验室克隆所得基因序列片段和 NCBI 上已公布的基因序列进行设计的，引物序列见表 2-3。

表 2-3 荧光定量引物

目标基因	正向引物（5′→3′）	反向引物（5′→3′）	退火温度/℃	登录号
CuZnSOD	CGCACTTCAACCCTTACA	ACTTTCCTCATTGCCTCC	61.5	GU901214
MnSOD	ACGACCCAAGTCTCCCTA	ACCCTGTGGTTCTCCTCC	60.4	GU218534
CAT	GAAGTTCTACACCGATGAGG	CCAGAAATCCCAAACCAT	58.7	FJ560431
GPx1a	GGGCTGGTTATTCTGGGC	AGGCGATGTCATTCCTGTTC	61.5	EU828796
GPx1b	TTTTGTCCTTGAAGTATGTCCGTC	GGGTCGTTCATAAAGGGCATT	60.3	KT757315
GPx4a	TACGCTGAGAGAGGTTTACACAT	CTTTTCCATTGGGTTGTTCC	60.4	KU255598
GPx4b	CTGGAGAAATACAGGGGTTACG	CTCCTGCTTTCCGAACTGGT	60.3	KU255599
GSTP1	ACAGTTGCCCAAGTTCCAG	CCTCACAGTCGTTTTTTCCA	59.3	KM112099
GSTP2	TGCCTTGAAGATTATGCTGG	GCTGGCTTTTATTTCACCCT	59.3	KP125490
GSTO1	GGTGCTCAATGCCAAGGGAA	CTCAAACGGGTCGGATGAA	58.4	KT757314
GSTO2	CTGCTCCCATCAGACCCATTT	TCTCCCCTTTTCTTGCCCATA	61.4	KU245630
GR	GTGTCCAACTTCTCCTGTG	ACTCTGGGGTCCAAAACG	59.4	JX854448
Nrf2	CTGGACGAGGAGACTGGA	ATCTGTGGTAGGTGGAAC	62.5	KF733814
Keap1a	TTCCACGCCCTCCTCAA	TGTACCTCCCGCTATG	63.0	KF811013
Keap1b	TCTGCTGTATGCGGTGGGC	CTCCTCCATTCATCTTTCTCG	57.9	KJ729125
caspase-2	CGCTGTTGTGTGTTTACTGTCTCA	ACGCCATTATCCATCTCCTCTC	60.3	KT757313
caspase-3	GCTGTGCTTCATTTGTTTG	TCTGAGATGTTATGGCTGTC	55.9	JQ793789
caspase-7	GCCATTACAGGATTGTTTCACC	CCTTATCTGTGCCATTGCGT	57.1	KT625601
caspase-8	ATCTGGTTGAAATCCGTGAA	TCCATCTGATGCCCATACAC	59.0	KM016991
caspase-9	CTGTGGCGGAGGTGAGAA	GTGCTGGAGGACATGGGAAT	59.0	JQ793787
Apaf-1	AAGTTCTGGAGCCTGGACAC	AACTCAAGACCCCACAGCAC	61.4	KM279717
Bax	CATCTATGAGCGGGTTCGTC	TTTATGGCTGGGGTCACACA	60.3	JQ793788.1
FasL	AGGAAATGCCCGCACAAATG	AACCGCTTTCATTGACCTGGAG	61.4	KT445873
Bcl-2	AGGAAAATGGAGGTTGGGAT	CTGAGCAAAAAGGCGATG	60.3	JQ713862.1
IAP	CACAATCCTGGTATGCGTCG	GGGTAATGCCTCTGGTGCTC	58.4	FJ593503.1
Mcl-1	TGGAAAGTCTCGTGGTAAAGCA	ATCGCTGAAGATTTCTGTTGCC	58.4	KT757307

续表

目标基因	正向引物（5′→3′）	反向引物（5′→3′）	退火温度/℃	登录号
p38 MAPK	TGGGAGCAGACCTCAACAAT	TACCATCGGGTGGCAACATA	60.4	KM112098
JNK	ACAGCGTAGATGTGGGTGATT	GCTCAAGGTTGTGGTCATACG	62.3	KT757312
ZO-1	CGGTGTCTTCGTAGTCGG	CAGTTGGTTTGGGTTTCAG	59.4	KJ000055
ZO-2	TACAGCGGGACTCTAAAATGG	TCACACGGTCGTTCTCAAAG	60.3	KM112095
occludin	TATCTGTATCACTACTGCGTCG	CATTCACCCAATCCTCCA	59.4	KF193855
claudin-b	GAGGGAATCTGGATGAGC	ATGGCAATGATGGTGAGA	57.0	KF193860
claudin-c	GAGGGAATCTGGATGAGC	CTGTTATGAAAGCGGCAC	59.4	KF193859
claudin-f	GCTGGAGTTGCCTGTCTTATTC	ACCAATCTCCCTCTTTTGTGTC	57.1	KM112097
claudin-7a	ACTTACCAGGGACTGTGGATGT	CACTATCATCAAAGCACGGGT	59.3	KT625604
claudin-7b	CTAACTGTGGTGGTGATGAC	AACAATGCTACAAAGGGCTG	59.3	KT445866
caudin-11	TCTCAACTGCTCTGTATCACTGC	TTTCTGGTTCACTTCCGAGG	62.3	KT445867
β-actin	GGCTGTGCTGTCCCTGTA	GGGCATAACCCTCGTAGAT	61.4	M25013

（b）总 RNA 提取。

具体步骤如下：①取 50 μg 的草鱼肠道组织（前、中或后肠），迅速放入装有 900 μL RNAiso Plus 的 2 mL EP 管中（含一大两小的钢珠），并经组织研磨仪（60 Hz）机打 30 s；②随后将 EP 管从组织研磨仪中取出，并置于冰上静止 3 min，随后在 EP 管中加入 300 μL 氯仿（4℃），并充分混匀，然后在冰上静止至分层；③将静置后的 EP 管置于冷冻离心机中（4℃，12000 g）离心 15 min；④同时取 300 μL 异丙醇于灭好菌的 1.5 mL EP 管中，并置于冰上预冷，随后从离好心的 EP 中吸取 150 μL 的上清液于该 1.5 mL EP 管中，上下反复颠倒，保证充分混匀，冰上静止 10 min，再离心 10 min（4 ℃，12000 g）；⑤随后在超净台中倒掉异丙醇，并用预冷好的 75%乙醇清洗两次，等乙醇挥发干后，加入适量预冷好的 DEPC 水稀释，并通过核酸蛋白检测仪测定其浓度和纯度。同时，通过琼脂糖凝胶电泳检测所提取的 RNA 完整性。

（c）cDNA 合成。

cDNA 的合成参考王开卓（2019）的方法进行。主要包括两个部分：①DNA 去除反应（5×gDNA Eraser Buffer，2 μL；gDNA Eraser，1 μL；总 RNA，2 μL；DEPC 水，5 μL；42℃，2 min；4℃，长期保存）；②反转录反应（5×PrimeScript

Buffer, 4 μL; PrimeScript RT Enzyme Mix I, 1 μL; RT Primer Mix, 1 μL; 上一步的反应液, 10 μL; DEPC 水, 4 μL) 37℃, 15 min; 85 ℃, 5 s; 4℃, 长期保存。待反应完成后, 将合成产物保存于-20℃备用。

(d) 荧光定量 PCR。

荧光定量 PCR 反应参考段绪东（2019）的方法进行。反应体系为 15 μL, 具体步骤如下：0.6 μL 正反向引物, 2 μL cDNA, 7.5 μL SYBR$^©$ Rremix Ex Taq^{TM}, 以及 4.3 μL 的 DEPC 水。qPCR 的程序为 30 s 95 ℃ 1 个循环；5 s 95 ℃, 30 s 退火温度（表 2-3）, 39 个循环。在检测草鱼肠道目标基因表达前, 参照 Su 等（2011）的方法, 我们采用 geNorm3.5 软件比较了常用管家基因（β-actin、GAPDH、18SrRNA、EF1-α）的稳定性, 根据预试验结果, β-actin 在本试验中是最理想的管家基因。根据溶解曲线, 确定 β-actin 和目标基因的扩增效率。在保证所有基因的扩增效率达到 100% 左右时, 参考 Livak 和 Schmittgen（2001）的方法, 采用 $2^{-\Delta\Delta CT}$ 法对目标基因表达进行定量分析。

2.1.6.10 蛋白表达

(1) 试剂和药品。

RIPA 强裂解液（碧云天, 中国）、PMSF（碧云天, 中国）、BCA 蛋白浓度测定试剂盒（碧云天, 中国）、30% Acr/Bic（北京索莱宝, 中国）、1.5 mol/L Tris-HCl（北京索莱宝, 中国）、1.0 mol/L Tris-HCl（北京索莱宝, 中国）、TEMED（碧云天, 中国）、滤纸（Bio-rad, 美国）、PVDF 膜（Millipore, 美国）、ECL 显影液（碧云天, 美国）、蛋白 marker（Bio-rad, 美国）、Tween-20（北京索莱宝, 中国）、Tris-Hcl（北京索莱宝, 中国）、甘氨酸（北京索莱宝, 中国）、BSA（碧云天, 中国）和 SDS-PAGE 蛋白上样缓冲液（5X）（碧云天, 中国）等。

(2) 仪器设备。

SDS-PAGE 电泳系统（Bio-rad, 美国）、湿转系统（Bio-rad, 美国）、SCIENTZ-48 高通量组织研磨仪（宁波新芝生物科技股份有限公司, 中国）、冷冻离心机（Eppendorf, 德国）、移液枪（Eppendorf, 德国）、凝胶成像分析仪（BIO-RAD, 美国）、TX223L 电子天平（SHIMADZU, 日本）和超净工作台 SB-2（上海净化设备厂, 中国）等。

(3) 试验方法。

(a) 样品制备。

①首先取约 50 μg 草鱼肠道组织（前、中或后肠）置于已灭菌的 2 mL EP 管（含 400 μL RIPA 和 5 μL PMSF）中，加入一大两小三颗钢珠，并经过组织研磨仪（50 Hz）机打 60 s；②随后将 EP 管置于冰上静置 30 min，以充分裂解组织；③离心 15 min（12000 r/min，4 ℃），随后吸取上清液，用 BCA 试剂盒（碧云天生物技术有限公司，上海）检测各芥酸处理组的上清液蛋白浓度，根据检测结果，用适当的裂解液将各管的上清液蛋白浓度调为一致；④在上清液中按 4∶1 的比例加入 5×蛋白上样缓冲液，充分混匀。变性并保存在 -80 ℃ 中备用。

（b）Western blotting 操作。

具体步骤如下：①将样品从 -80 ℃ 冰箱中取出，并变性；②按 40 μg 蛋白/泳道的量取适量上清液，经 10% 丙烯酰胺凝胶电泳进行分离；③将分离后的蛋白通过湿转（恒压 100 v，90 min）转移到 PVDF 膜上；④将完成湿转后的 PVDF 膜用 5% 的 BSA 封闭 90 min；⑤4 ℃ 条件下孵育一抗过夜（约 13 h）；⑥二抗孵育 2 h，ECL 发光剂（Affinity Biologicals Inc.，Canada）进行显影，用 Image Lab™（Bio-Rad Laboratories, Inc.）软件进行凝胶成像，使用 Image J 1.38 软件对蛋白条带进行定量分析。参考本实验室之前的研究，选择 β-actin 蛋白作为内参蛋白。一抗：Nrf2（1∶1000 稀释，Abcam，英国）、E-cadherin（1∶1000，ABclonal，中国）、β-catenin（1∶1000，ABclonal，中国）、ZO-1（1∶1000 稀释，ABclonal，中国）；La min B1（1∶1000，Affinity BioReagents，美国）和 β-actin（1∶3000，Affinity BioReagents，美国）。二抗：辣根过氧化物酶标记山羊抗兔 IgG（H+L）（1∶8000，碧云天，中国）。

2.1.6.11 RhoA 活性检测

（1）试剂和药品。

RhoA 活性检测试剂盒购于 Cytoskeleton（丹佛，美国）。其他试剂和药品同上。

（2）仪器设备。

同上。

（3）试验方法。

RhoA 活性检测主要参考 Ren 等（1999）的方法进行，主要操作如下：①取 50 μg 草鱼肠道组织置于已灭菌的 2 mL EP 管（含 400 μL RIPA 和 5 μL PMSF）中，加入一大两小三颗钢珠，并经过组织研磨仪（50 Hz）机打 60 s；②将含有肠道组织裂解液的 EP 管置于离心机中离心 3 min（14000 r/min，4 ℃）；③取等量的上清液与 GST-RBD 珠（20 μg）在 4℃ 条件下孵育 1 h；④孵育结束后，用洗涤缓冲液洗 GST-RBD 珠 4 次，结合的 RhoA 蛋白通过 western blot 的方法检测。同时，取等量的上清液通过 western blot 检测总 RhoA 蛋白的含量。RhoA 的活性通过结合的 RhoA 量与总 RhoA 量的比值确定。Western blot 操作方法同上。

2.1.7 统计分析

本试验所有数据均采用 SAS 软件（SAS Institute, Inc., 2006）的一般线性模型程序进行分析。$P \leq 0.05$ 为差异显著。根据 Hooft 等（2011）的方法，采用正交多项对比分析芥酸的一次和二次效应。根据 Robbins 等（2006）的方法，采用折线回归分析（SAS），计算生长中期草鱼饲料中芥酸的控制剂量。数据相关性采用 SAS 软件的 PROC CORR 程序进行分析。

2.2 试验结果

2.2.1 芥酸对生长中期草鱼生产性能的影响

芥酸对生长中期草鱼生产性能的影响如表 2-4 所示。随着饲料中芥酸水平的升高，生长中期草鱼的终末体重、增重百分比、特定生长率、摄食量、饲料效率、肠重和肠体指数呈现出线性（$P<0.01$）和二次（$P<0.05$）降低的变化，以及肠长和肠长指数呈现出线性（$P<0.01$）降低的变化。生长中期草鱼前肠中的芥酸残留量随着饲料芥酸水平的升高呈现出线性和二次升高的变化（$P<0.01$），同时，中肠和后肠的芥酸残留量随着饲料芥酸水平的升高呈现出线性升高的变化（$P<0.01$）。

表 2-4 芥酸对生长中期草鱼生长、肠道芥酸残留（μg/g 组织）和肠道生长的影响

测量指标	饲料芥酸水平/（% diet）						SEM	P-值	
	0.00（对照）	0.29	0.60	0.88	1.21	1.50		线性	二次
初重/g	128.56	129.44	129.22	129.67	129.00	129.11	0.83	0.66	0.22
末重/g	471.15	468.97	466.40	417.19	311.52	261.86	7.12	<0.01	<0.01
PWG/%	267.56	263.32	261.97	222.78	142.53	103.84	4.92	<0.01	<0.01

续表

测量指标	饲料芥酸水平/（% diet）						SEM	P-值	
	0.00（对照）	0.29	0.60	0.88	1.21	1.50		线性	二次
SGR/（%·d^{-1}）	2.16	2.15	2.14	1.95	1.47	1.18	0.03	<0.01	<0.01
FI/（g·fish^{-1}）	476.86	471.19	464.11	431.41	294.35	228.76	7.23	<0.01	<0.01
FE/%	0.72	0.72	0.73	0.67	0.62	0.59	0.01	<0.01	<0.01
IL/cm	64.23	63.13	62.03	57.18	54.64	52.53	2.39	<0.01	0.11
ILI/%	194.30	191.65	188.25	180.71	175.64	171.29	8.05	<0.01	0.52
IW/g	14.76	14.36	13.95	11.16	9.31	8.37	0.58	<0.01	<0.01
ISI/%	3.09	3.06	3.02	2.88	2.67	2.56	0.15	<0.01	<0.05
芥酸含量/（μg/g组织）									
PI	12.94	29.54	71.21	107.45	193.96	364.92	31.28	<0.01	<0.01
MI	17.22	74.82	87.33	168.91	176.74	229.09	9.95	<0.01	0.29
DI	11.42	67.78	79.40	129.60	151.39	217.99	17.95	<0.01	0.41

注　PWG（%）：增重百分比；SGR（%/d）：特定生长率；FI（g/尾）：饲料摄入；FE（%）：饲料效率。P值表示呈现出明显的线性和二次剂量关系（$P<0.01$）。

芥酸对生长中期草鱼饲料摄入量的影响如图2-1所示。在养殖的第0至第10天，各芥酸处理组草鱼的饲料摄入量都会逐渐增加，并且增加趋势比较接近。在第10天至第50天，1.21%和1.50%芥酸处理组草鱼的饲料摄入量低于其他各组。第50天至第60天，0.88%、1.21%和1.50%芥酸处理组草鱼的饲料摄入量低于其他各组。在整个养殖期间，0.00、0.29和0.88%芥酸处理组草鱼的饲料摄入量变化趋势几乎一致。

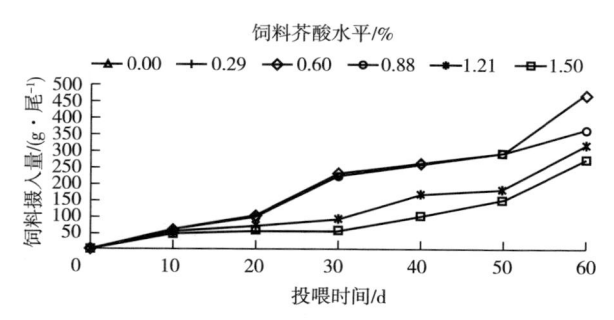

图2-1　芥酸对生长中期草鱼饲料摄入量的影响

2.2.2 芥酸对生长中期草鱼营养物质表观消化率的影响

芥酸对草鱼营养物质表观消化率的影响如图 2-2 所示。与对照组相比，饲料中的芥酸会明显降低生长中期草鱼饲料干物质、粗脂肪和粗蛋白的表观消化率（$P<0.01$）。

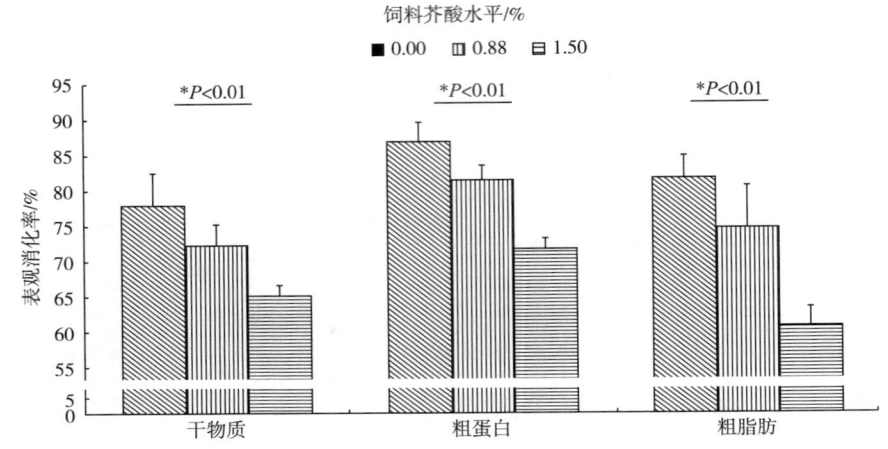

图 2-2 芥酸对生长中期草鱼营养物质表观消化率的影响

数据表示为平均值±标准误（$n=3$，30条鱼/组）；实线上方的 P 值表示呈现出明显的线性剂量关系（$P<0.01$）

2.2.3 芥酸对生长中期草鱼肠道表观症状和组织结构的影响

芥酸对生长中期草鱼肠道表观症状和组织结构的影响如图 2-3 和图 2-4 所示。对照组中生长中期草鱼肠道外观呈肉色，颜色正常，并且前、中和后肠绒毛结构完整，无增生现象。与对照组相比，饲料中芥酸（≥0.88%）可引起生长中期草鱼肠道充血以及前、中和后肠绒毛组织增生。

2.2.4 芥酸对生长中期草鱼血清二胺氧化酶活性和 D-乳酸含量的影响

芥酸对生长中期草鱼血清二胺氧化酶活性和 D-乳酸含量的影响如表 2-5 所示。随着饲料中芥酸水平的升高，生长中期草鱼血清二胺氧化酶活性呈现出线性（$P<0.01$）升高的变化，血清 D-乳酸含量呈现出线性（$P<0.01$）和二次（$P<0.01$）升高的变化。

第二章 芥酸对生长中期草鱼生产性能和肠道结构完整性的影响

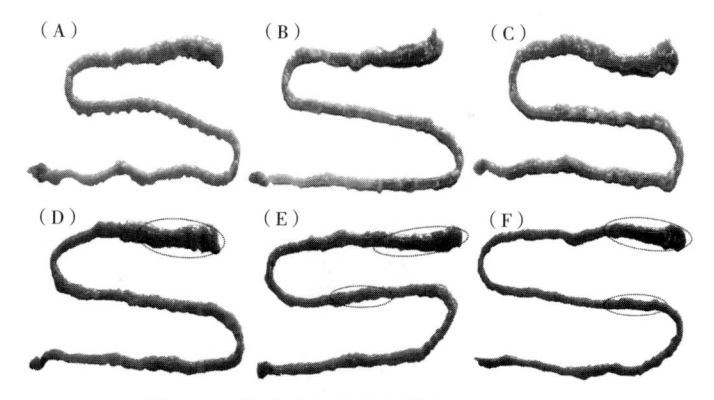

图 2-3 芥酸对生长中期草鱼肠道的影响

（A）对照组　（B）0.29% 芥酸组　（C）0.60% 芥酸组　（D）0.88% 芥酸组　（E）1.21% 芥酸组　（F）1.50% 芥酸组　在各图中，椭圆形区域表示草鱼肠道出现充血症状

前肠

中肠

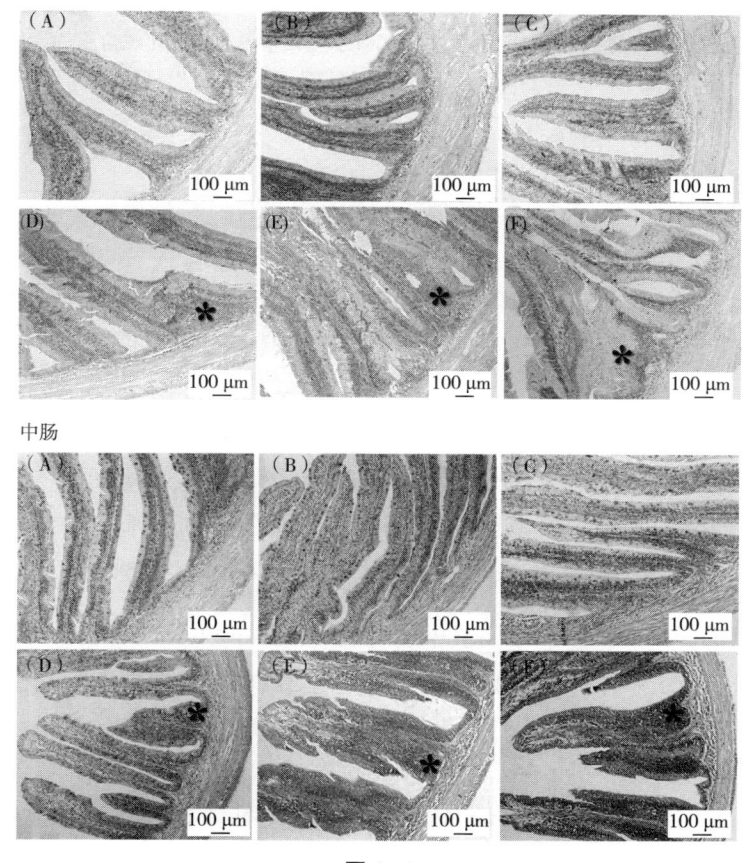

图 2-4

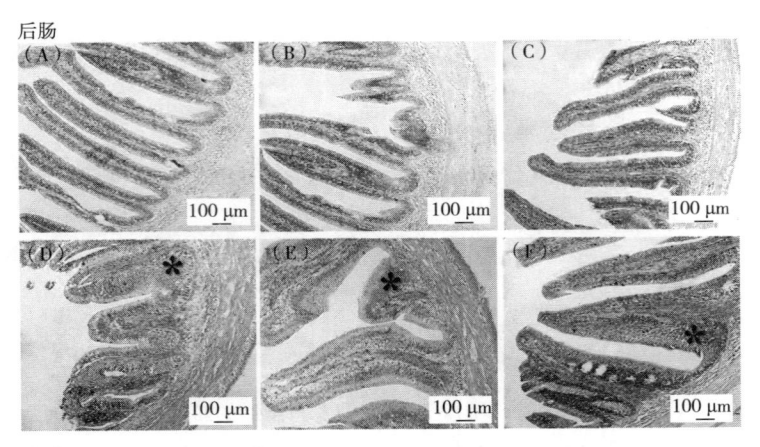

图 2-4　芥酸对生长中期草鱼前肠、中肠和后肠组织结构的影响（100×）

（A）对照组　（B）0.29% 芥酸组　（C）0.60% 芥酸组　（D）0.88% 芥酸组
（E）1.21%芥酸组　（F）1.50%芥酸组　在各图中，*表示草鱼各肠段组织出现绒毛增生的症状

表 2-5　芥酸对生长中期草鱼血清二胺氧化酶活性和 D-乳酸含量的影响

测量指标	饲料芥酸水平/（%饲料）						SEM	P-值	
	0.00（对照）	0.29	0.60	0.88	1.21	1.50		线性	二次
二胺氧化酶/（U·L^{-1}）	446.48	476.72	491.72	560.95	599.48	667.43	73.67	<0.01	0.34
D-乳酸/（μmol·L^{-1}）	34.82	36.80	39.66	51.72	63.64	79.37	4.62	<0.01	<0.01

2.2.5　芥酸影响生长中期草鱼肠道结构完整性的可能机制

2.2.5.1　芥酸对生长中期草鱼肠道细胞结构完整性的影响

芥酸对生长中期草鱼前、中和后肠氧化损伤的影响如表 2-6 所示。随着饲料中芥酸水平的升高，草鱼三个肠段中的 ROS、MDA 和 PC 呈现出线性（$P<0.01$）和二次（$P<0.05$）升高的变化。随着饲料中芥酸水平的升高，草鱼前肠中的 AHR、MnSOD、GPX 和 GST 的活性，中肠中的 AHR、ASA、MnSOD、CAT、GST 和 GR 活性以及后肠中的 GSH 含量；MnSOD、GPX、GST 和 GR 活性呈现出线性（$P<0.01$）和二次（$P<0.05$）降低的变化。随着饲料中芥酸水平的升高，前肠中的 GSH 含量，ASA、CAT 和 GR 活性，中肠中的 GSH 含量和 GPX 活性，以及

后肠中的 AHR、ASA 和 CAT 活性呈现出线性（$P<0.01$）降低的变化。有趣的是，饲料中芥酸水平对生长中期草鱼前、中和后肠中 CuZnSOD 活性没有影响（$P>0.05$）。

表2-6 芥酸对生长中期草鱼前、中和后肠抗氧化指标的影响

肠道组织	饲料芥酸水平/（%饲料）						SEM	P-值	
	0.00（对照）	0.29	0.60	0.88	1.21	1.50		线性	二次
前肠（PI）									
ROS	100.00	102.85	109.96	123.18	142.30	170.48	9.59	<0.01	<0.01
MDA	31.11	32.90	34.04	39.80	43.37	50.63	2.96	<0.01	<0.05
PC	7.25	7.32	7.83	9.92	11.73	13.33	0.72	<0.01	<0.01
AHR	67.14	66.87	64.53	58.27	56.20	45.78	4.88	<0.01	<0.05
ASA	139.90	137.76	130.23	122.56	109.14	99.41	10.29	<0.01	0.13
CuZnSOD	5.09	4.99	4.96	4.88	4.85	4.82	0.48	0.26	0.85
MnSOD	7.40	7.38	7.29	6.32	5.70	4.75	0.31	<0.01	<0.01
CAT	3.60	3.51	3.17	3.12	2.93	2.74	0.233	<0.01	0.83
GPx	175.63	168.61	164.01	141.33	129.10	107.82	10.63	<0.01	<0.05
GST	128.98	128.88	119.32	101.22	86.39	77.06	7.39	<0.01	<0.05
GR	54.39	53.46	42.85	42.71	38.65	33.12	4.05	<0.01	0.75
GSH	6.69	6.66	6.38	5.89	5.66	5.21	0.48	<0.01	0.32
中肠（MI）									
ROS	100.00	104.21	107.65	128.83	133.17	161.56	7.31	<0.01	<0.01
MDA	36.55	37.58	40.07	48.88	51.99	58.75	2.86	<0.01	<0.05
PC	6.11	6.56	6.69	7.24	8.29	10.77	0.52	<0.01	<0.01
AHR	57.31	56.91	52.96	47.05	41.95	32.43	3.90	<0.01	<0.05
ASA	148.06	142.62	136.68	111.52	105.34	82.55	9.63	<0.01	<0.05
CuZnSOD	6.88	6.67	6.65	6.55	6.53	6.49	0.39	0.07	0.55
MnSOD	6.42	6.49	6.43	5.91	5.19	4.52	0.38	<0.01	<0.01
CAT	2.39	2.31	2.26	2.10	1.91	1.65	0.14	<0.01	<0.05
GPx	103.99	100.54	93.74	84.52	78.63	64.61	6.58	<0.01	0.07
GST	102.07	100.41	99.07	81.84	70.78	57.81	6.60	<0.01	<0.01
GR	46.31	46.20	42.39	36.70	31.73	26.59	3.32	<0.01	<0.05
GSH	5.11	5.09	4.70	4.36	3.92	3.52	0.33	<0.01	0.13

续表

肠道组织	饲料芥酸水平/（%饲料）						SEM	P-值	
	0.00（对照）	0.29	0.60	0.88	1.21	1.50		线性	二次
后肠（DI）									
ROS	99.99	101.48	108.45	120.36	130.02	153.68	8.77	<0.01	<0.05
MDA	40.41	40.98	42.20	47.14	52.67	59.95	2.24	<0.01	<0.01
PC	5.81	5.98	6.16	7.24	8.41	9.60	0.62	<0.01	<0.01
AHR	51.76	51.66	45.73	40.45	38.18	29.18	4.03	<0.01	0.07
ASA	121.85	106.34	102.55	97.75	89.38	80.84	8.12	<0.01	0.56
CuZnSOD	6.02	6.00	5.97	5.91	5.88	5.82	0.48	0.40	0.92
MnSOD	8.46	8.37	8.38	7.60	6.42	5.53	0.52	<0.01	<0.01
CAT	2.31	2.29	2.21	1.92	1.70	1.66	0.19	<0.01	0.42
GPx	111.60	109.52	105.37	91.20	78.37	65.30	6.90	<0.01	<0.05
GST	79.54	78.70	71.23	70.32	56.99	40.22	5.84	<0.01	<0.01
GR	38.39	37.71	36.87	26.19	22.26	13.44	1.97	<0.01	<0.01
GSH	6.98	6.91	6.50	5.88	4.98	3.64	0.41	<0.01	<0.01

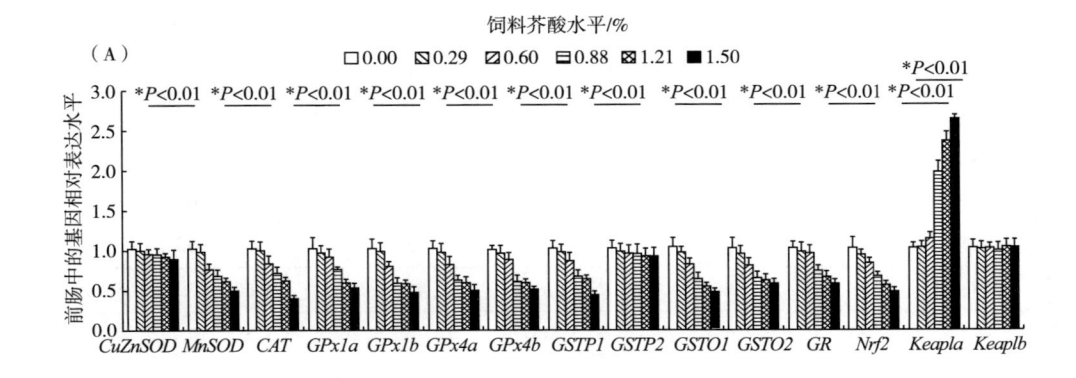

芥酸对生长中期草鱼前、中和后肠抗氧化酶相关基因 mRNA 水平的影响如图 2-5 所示。随着饲料中芥酸水平的升高，草鱼三个肠段中 *MnSOD*、*CAT*、*GPX1a*、*GPX1b*、*GPX4a*、*GPX4b*、*GSTP1*、*GSTO1*、*GSTO2*、*GR* 和 *Nrf2* 的 mRNA 水平呈现出线性降低的变化（$P<0.01$），而 *keap1a* 的 mRNA 水平呈现出线性升高的变化（$P<0.01$）。有趣的是，饲料中芥酸水平对生长中期草鱼前、中和后肠中 *CuZnSOD*、*GSTP2* 和 *keap1b* 的 mRNA 水平没有影响（$P>0.05$）。

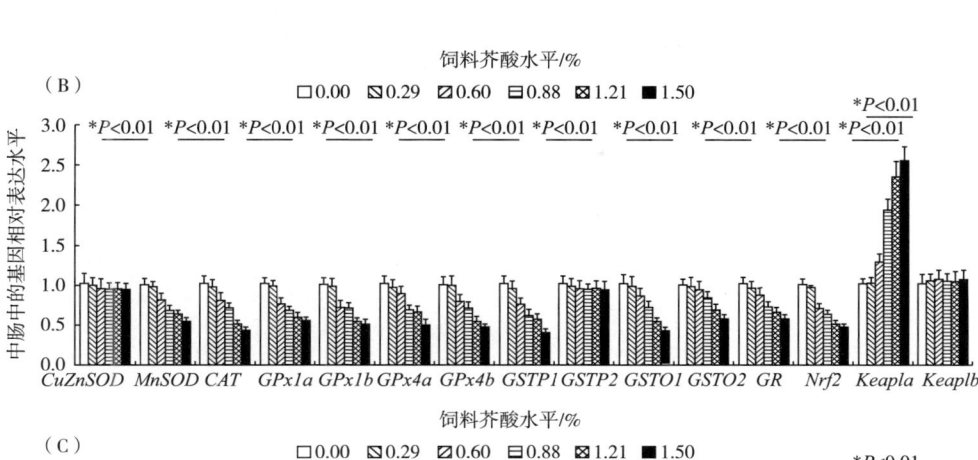

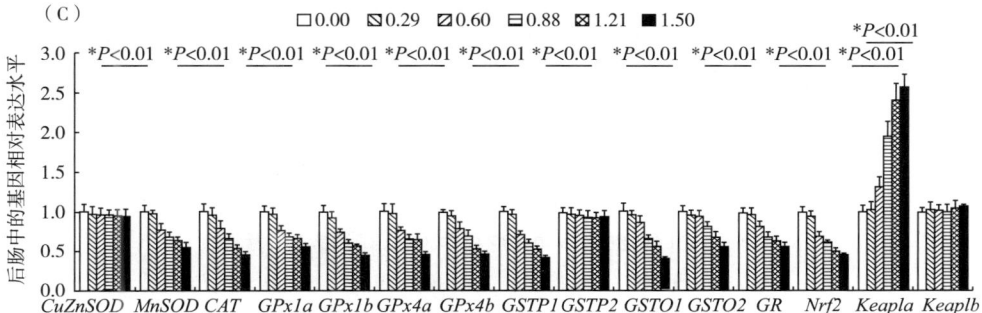

图 2-5 芥酸对生长中期草鱼前肠（A）、中肠（B）和
后肠（C）抗氧化酶相关基因 mRNA 水平的影响

数据表示为平均值±标准误（$n=3$，6 条鱼/组）；实线上方的 P 值表示
呈现出明显的线性剂量关系（$P<0.01$）

芥酸对生长中期草鱼前、中和后肠中细胞质和细胞核 Nrf2 蛋白质表达的影响如图 2-6 所示。随着饲料中芥酸水平的升高，草鱼三个肠段中细胞质和细胞核 Nrf2 的蛋白质表达呈现出线性降低的变化（$P<0.01$）。

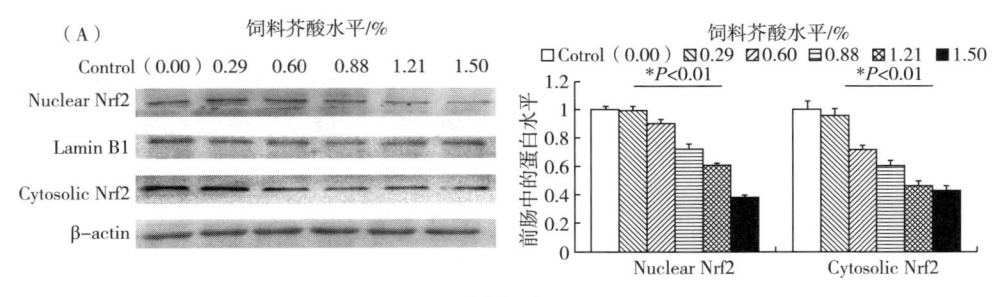

图 2-6

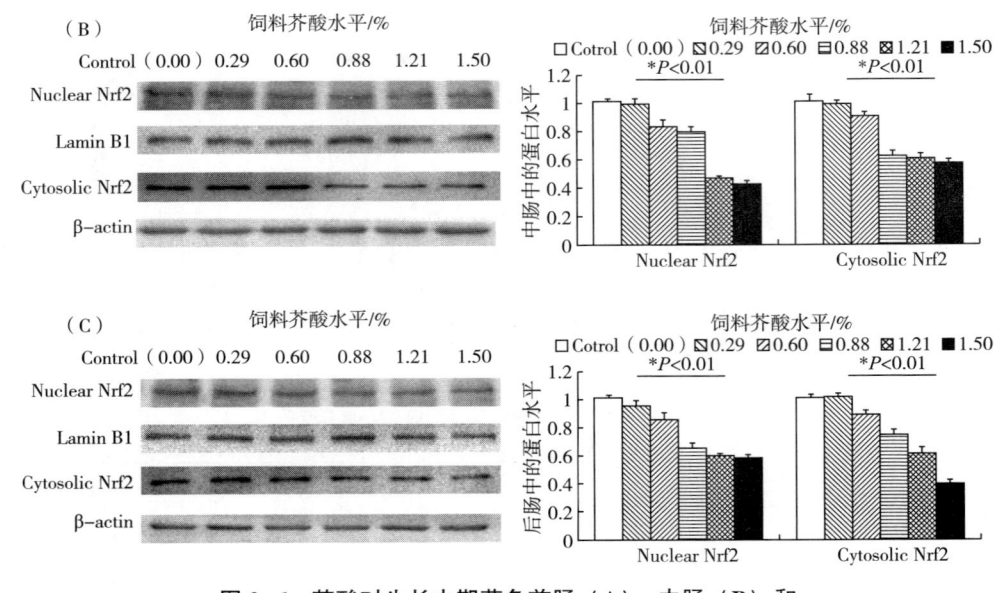

图2-6 芥酸对生长中期草鱼前肠（A）、中肠（B）和后肠（C）中细胞质和细胞核中 Nrf2 蛋白表达的影响

数据表示为平均值±标准误（$n=3$，6条鱼/组）；实线上方的 P 值表示呈现出明显的线性剂量关系（$P<0.01$）

芥酸对生长中期前、中和后肠 DNA 片段化的影响如图2-7所示。当饲料中芥酸水平达到0.88%或更高时，草鱼三个肠段都会出现 DNA 片段化现象。

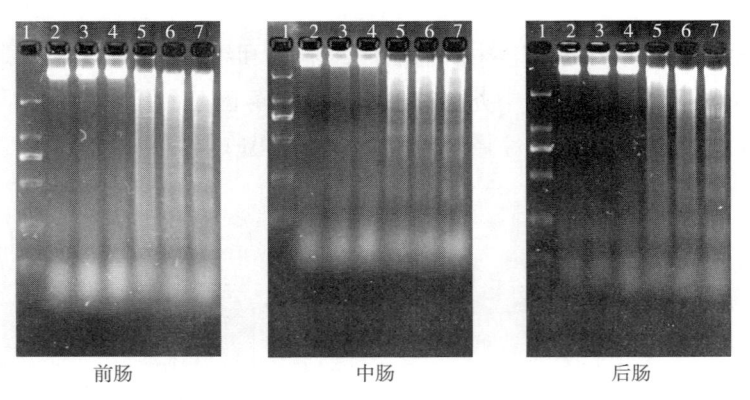

图2-7 芥酸对生长中期草鱼前肠、中肠和后肠中 DNA 片段化的影响

泳道1：Marker；泳道2-7：对照组、0.29% 芥酸组、0.60% 芥酸组、0.88% 芥酸组、1.21% 芥酸组和1.50% 芥酸组

芥酸对生长中期草鱼前、中和后肠凋亡相关基因 mRNA 水平的影响如图 2-8 所示。随着饲料中芥酸水平的升高,草鱼三个肠段中 caspase-2、caspase-7、caspase-8、caspase-9、Apaf-1、Bax、FasL 和 p38MAPK 的 mRNA 水平呈现出线性

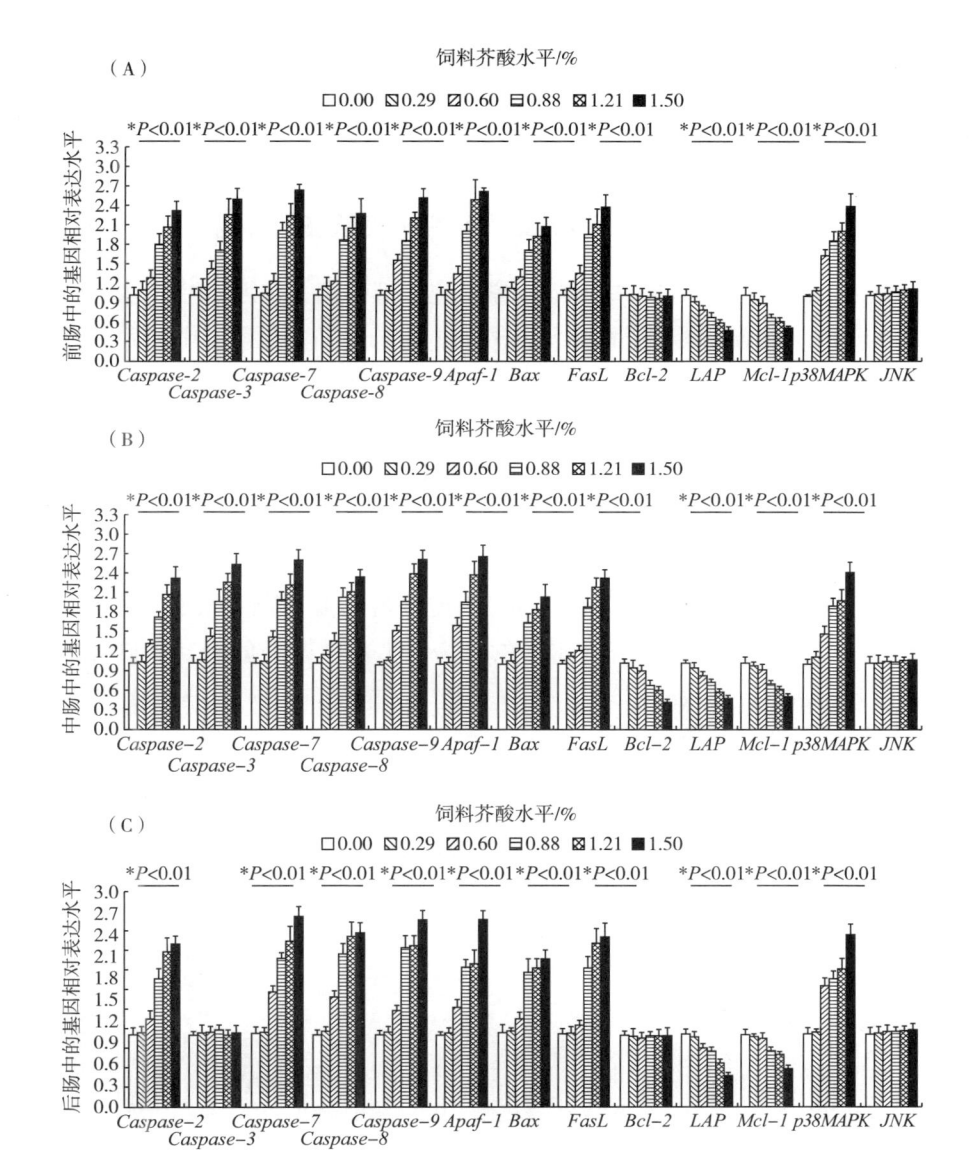

图 2-8 芥酸对生长中期草鱼前肠(A)、中肠(B)和
后肠(C)凋亡相关基因 mRNA 水平的影响

数据表示为平均值±标准误($n=3$,6 条鱼/组);实线上方的 P 值
表示呈现出明显的线性剂量关系($P<0.01$)

升高的变化（$P<0.01$），而 *IAP* 和 *Mcl-1* 的 mRNA 水平呈现出线性降低的变化（$P<0.01$）。有趣的是，随着饲料中芥酸水平的升高，前肠和中肠 *caspase-3* 的 mRNA 水平呈现出线性升高的变化（$P<0.01$），但后肠 *caspase-3* 的 mRNA 水平没有变化（$P>0.05$）。随着饲料中芥酸水平的升高，中肠 *Bcl-2* 的 mRNA 水平呈现出线性降低的变化（$P<0.01$），而前肠和后肠 *Bcl-2* 的 mRNA 水平没有变化（$P>0.05$）。饲料中芥酸水平对生长中期草鱼前、中和后肠中 *JNK* 的 mRNA 水平没有影响（$P>0.05$）。

2.2.5.2 芥酸对生长中期草鱼肠道细胞间结构完整性的影响

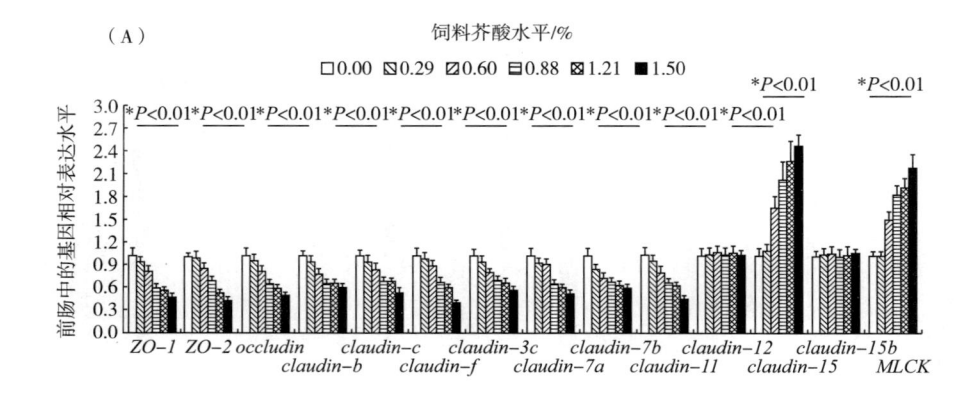

芥酸对生长中期草鱼前、中和后肠紧密连接蛋白相关基因 mRNA 水平的影响如图 2-9 所示。随着饲料中芥酸水平的升高，生长中期草鱼前、中和后肠紧密连接蛋白 *ZO-1*、*ZO-2*、*occludin*、*claudin-b*、*claudin-c*、*claudin-f*、*claudin-3c*、*claudin-7a*、*claudin-7b* 和 *claudin-11* 的 mRNA 水平呈现出线性降低的变化（$P<0.01$）以及 *claudin-15a* 的 mRNA 水平呈现出线性升高的变化（$P<0.01$）。有趣的是，饲料中芥酸水平对生长中期草鱼前、中和后肠中 *claudin-12* 和 *claudin-15b* 的 mRNA 水平没有影响（$P>0.05$）。

芥酸对生长中期草鱼前、中和后肠黏附连接蛋白相关基因 mRNA 水平的影响如图 2-10 所示。随着饲料中芥酸水平的升高，生长中期草鱼前、中和后肠黏附连接蛋白 *α-catenin*、*β-catenin*、*E-cadherin*、*nectin* 和 *afadin* 的 mRNA 水平呈现出线性降低的变化（$P<0.01$）。

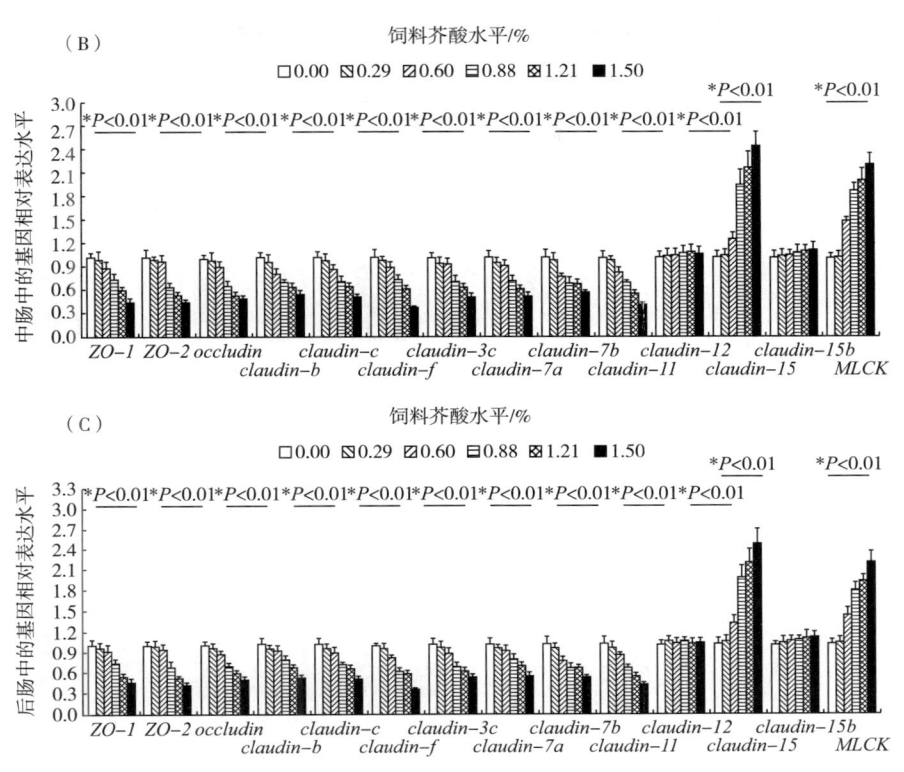

图 2-9 芥酸对生长中期草鱼前肠（A）、中肠（B）和后肠（C）紧密连接蛋白相关基因 mRNA 水平的影响

数据表示为平均值±标准误（$n=3$，6 条鱼/组）；实线上方的 P 值表示呈现出明显的线性剂量关系（$P<0.01$）

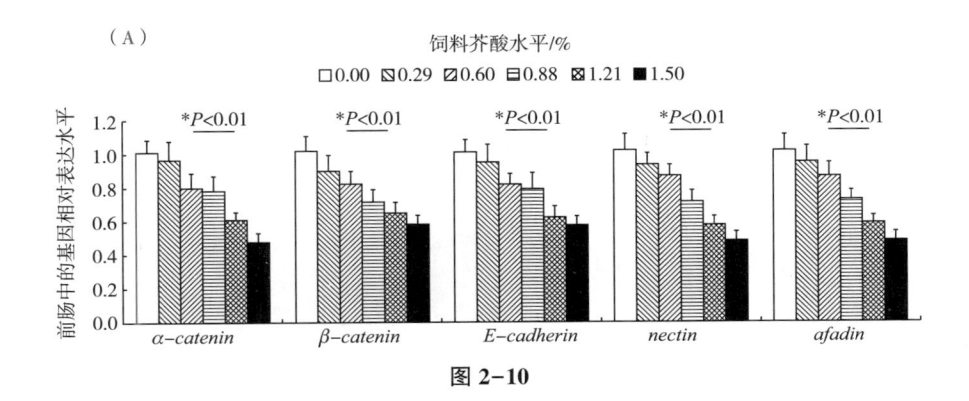

图 2-10

(B) 饲料芥酸水平/%
□0.00 ◨0.29 ◪0.60 ☰0.88 ▦1.21 ■1.50

(C) 饲料芥酸水平/%
□0.00 ◨0.29 ◪0.60 ☰0.88 ▦1.21 ■1.50

图 2-10　芥酸对生长中期草鱼前肠（A）、中肠（B）和
后肠（C）黏附连接蛋白相关基因 mRNA 水平的影响

数据表示为平均值±标准误（$n=3$，6 条鱼/组）；实线上方的 P 值
表示呈现出明显的线性剂量关系（$P<0.01$）

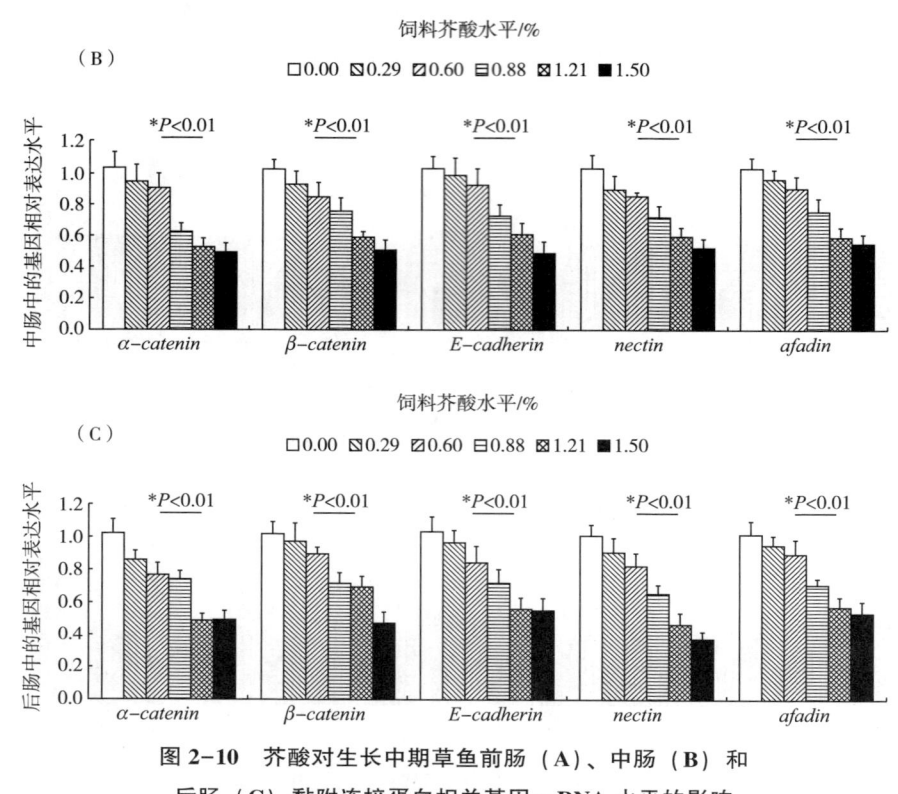

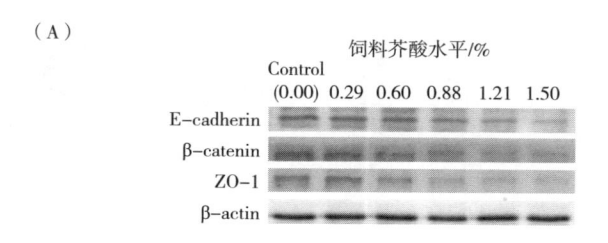

芥酸对生长中期草鱼前、中和后肠 E-cadherin、β-catenin 和 ZO-1 蛋白表达的影响如图 2-11 所示。随着饲料中芥酸水平的升高，生长中期草鱼前、中和后肠 E-cadherin、β-catenin 和 ZO-1 的蛋白表达水平呈现出线性降低的变化（$P<0.01$）。

(A) 饲料芥酸水平/%
Control (0.00)　0.29　0.60　0.88　1.21　1.50

E-cadherin
β-catenin
ZO-1
β-actin

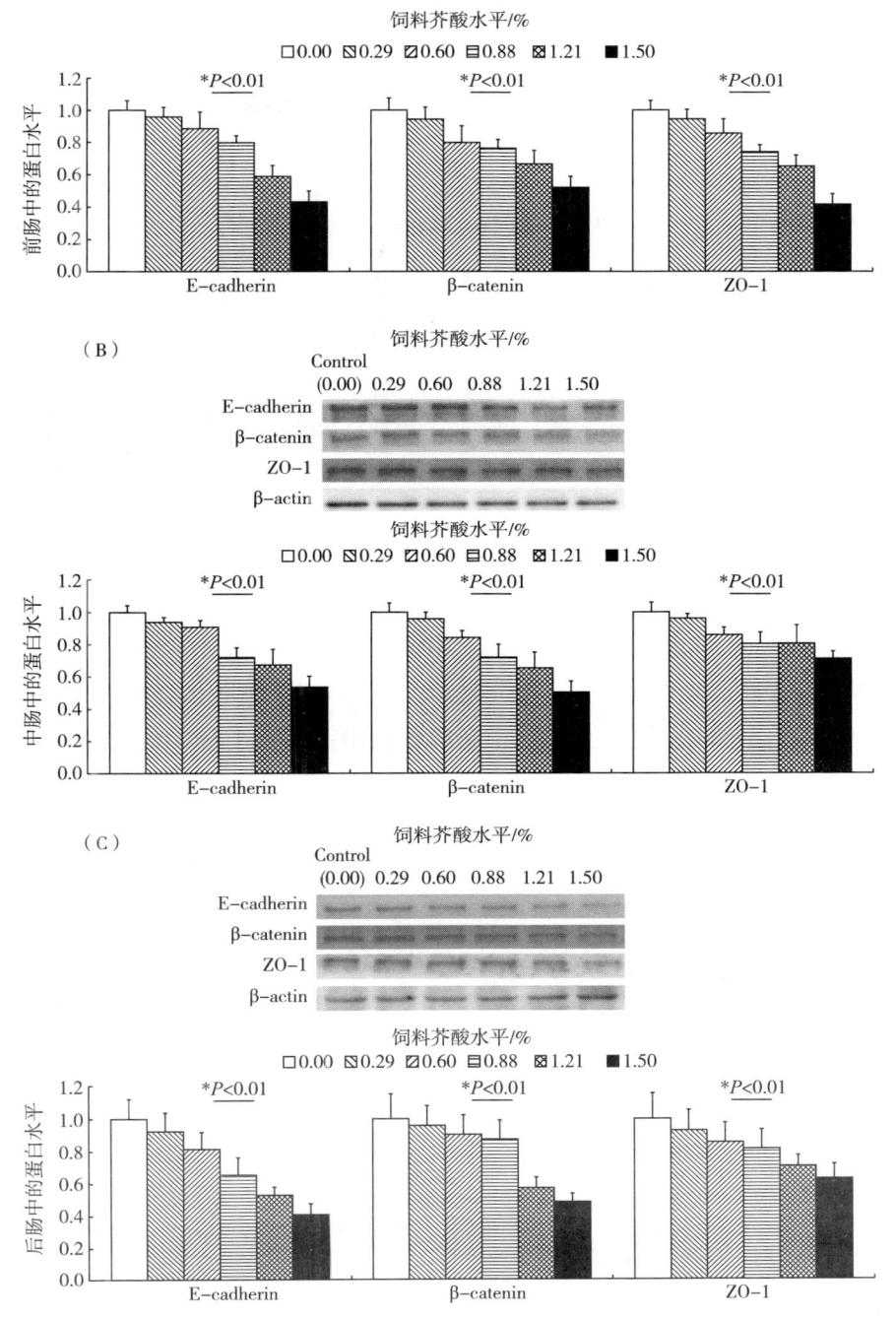

图 2-11 芥酸对生长中期草鱼前肠（A）、中肠（B）和后肠（C）
E-cadherin、β-catenin 和 ZO-1 蛋白表达的影响

数据表示为平均值±标准误（$n=3$，6 条鱼/组）；实线上方的 P 值
表示呈现出明显的线性剂量关系（$P<0.01$）

芥酸对生长中期草鱼前、中和后肠 NMⅡ、ROCK 和 MLCK 的 mRNA 水平的影响如图 2-12 所示。随着饲料中芥酸水平的升高，草鱼前、中和后肠关键信号分子 NMⅡ、ROCK 和 MLCK 的 mRNA 水平呈现出线性升高的变化（$P<0.01$）。

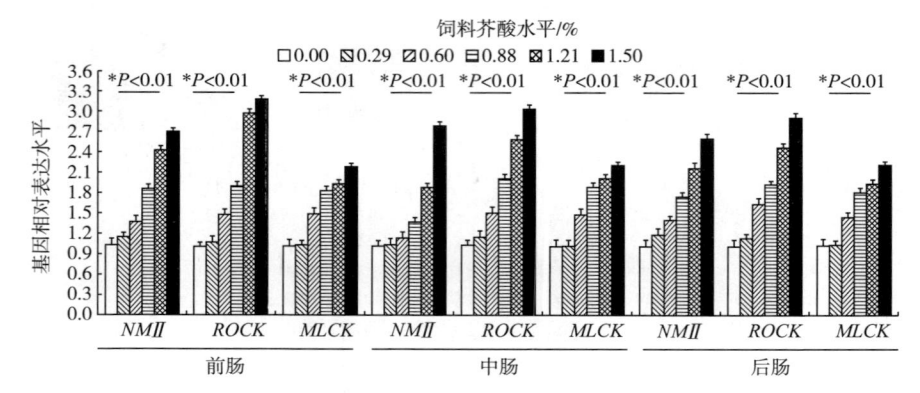

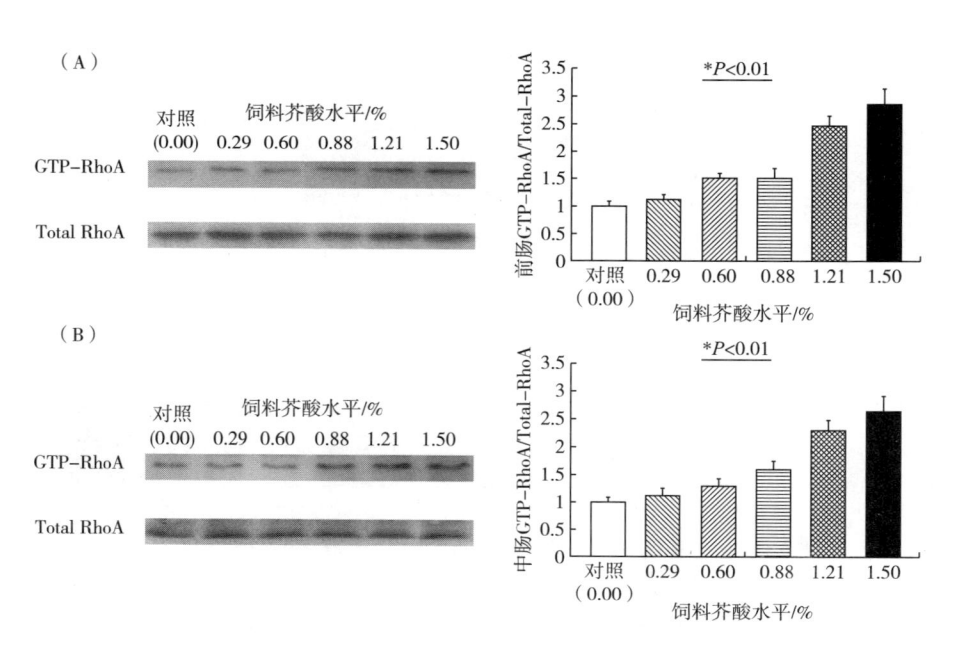

图 2-12　芥酸对生长中期草鱼前、中和后肠关键信号分子 mRNA 水平的影响

数据表示为平均值±标准误（$n=3$，6 条鱼/组）；实线上方的 P 值
表示呈现出明显的线性剂量关系（$P<0.01$）

芥酸对生长中期草鱼前、中和后肠 RhoA 活性的影响如图 2-13 所示。随着饲料中芥酸水平的升高，草鱼前、中和后肠 GTP-RhoA/Total-RhoA 比值呈现出线性升高的变化（$P<0.01$）。

（C）

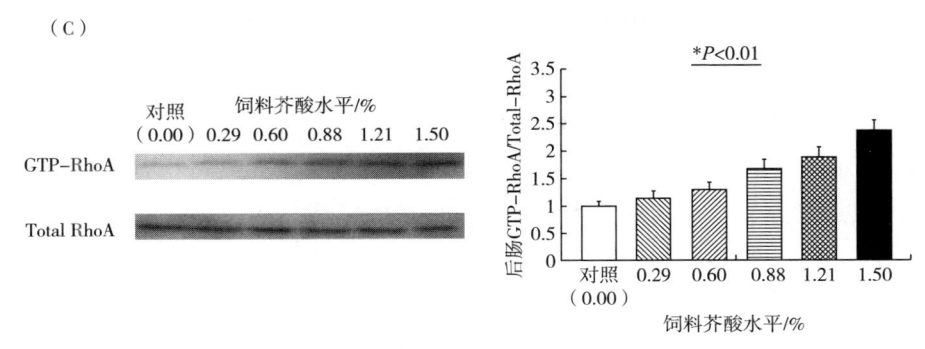

图 2-13 芥酸对生长中期草鱼前肠（A）、中肠（B）和
后肠（C）中 GTP-RhoA/Total-RhoA 比值的影响

数据表示为平均值±标准误（$n=3$，6条鱼/组）；实线上方的 P 值
表示呈现出明显的线性剂量关系（$P<0.01$）

2.3 讨论

2.3.1 芥酸降低生长中期草鱼的生产性能

本试验的研究结果表明：芥酸（≥0.64%）会降低生长中期草鱼的生产性能，这说明芥酸会对鱼类的生长产生负面影响。芥酸抑制动物生长在陆生动物的试验中也有报道。例如，5 g/kg 的芥酸会抑制小鸡的生长；4 g/kg 的芥酸会抑制羔羊的生长；22.3 g/kg 的芥酸会抑制猪的生长以及 2.2 g/kg 的芥酸会抑制大鼠的生长。由此可见，芥酸会抑制动物的生长，但在不同的物种上，芥酸的毒性剂量可能会存在差异。此外，菜籽粕中其他抗营养因子也会抑制鱼类的生长，如硫代葡萄糖苷、植酸和单宁等。据报道，动物生长依赖于其对饲料中营养物质的利用能力。因此，我们接下来研究了芥酸对草鱼营养物质表观消化率的影响。

2.3.2 芥酸降低生长中期草鱼营养物质表观消化率

本试验的研究结果表明：芥酸会降低生长中期草鱼粗蛋白、粗脂肪和干物质表观消化率。陆生动物上的研究也发现，芥酸会降低羊羔粗蛋白、粗脂肪和干物质的表观消化率以及降低小鸡脂肪酸的表观消化率，这与本试验的研究结果一致。与此同时，有关抗营养因子降低鱼类营养物质利用能力的现象在其他研究中也有报道。在大菱鲆上的研究发现，大豆球蛋白和β-伴大豆球蛋白都会降低其干物质和粗蛋白的表观消化率。同时，在大西洋鲑上的研究表明，植酸和大豆皂甙都会降低其粗脂肪的表观消化率。这些研究结果说明：抗营养因子抑制鱼类产生性能可能跟其降低鱼类营养物质表观消化率有关。动物肠道是动物营养物质消化吸收的主要器官，动物营养物质表观消化率的降低与动物肠道发育密切相关。肠体指数和肠长指数是肠道发育的重要标识。在本试验中，芥酸会降低生长中期草鱼的肠体指数和肠长指数，表明芥酸会抑制草鱼肠道的发育。此外，在小鼠上的研究表明，芥酸主要在肠道蓄积。本试验的研究结果发现，芥酸会在草鱼肠道中大量蓄积，并且随着饲料中芥酸含量的增多，草鱼肠道中芥酸的残留量会明显增多。因此，我们猜测：芥酸在草鱼肠道中蓄积进而对肠道产生负面影响可能是芥酸降低草鱼营养物质利用能力和抑制其生长的原因之一。围绕这一假设，我们接下来研究芥酸对草鱼肠道的影响。

2.3.3 芥酸破坏生长中期草鱼肠道结构完整性

肠道结构受损是毒性物质对动物肠道产生负面影响的主要方式，其可通过肠道充血和肠道绒毛组织增生来进行评估。在本试验中，芥酸会导致生长中期草鱼肠道充血和前、中、后肠三个肠段的肠道绒毛组织增生，说明芥酸会破坏草鱼肠道结构完整性。抗营养因子上的相关研究表明：β-伴大豆球蛋白会改变草鱼肠道结构（抑制肠道发育），导致金鲫鱼肠道绒毛的脱落和部分上皮细胞与固有层发生分离，以及引起大菱鲆肠道中黏膜褶皱的数量减少和高度变短以及肠褶皱内固有层变宽；大豆皂甙会导致大菱鲆肠道绒毛高度变短、肠道绒毛间发生融合以及肠道固有层和黏膜下层变宽；棉酚会导致生长中期草鱼肠道上皮细胞排列疏松和上皮脱落；植酸会导致生长中期草鱼肠道固有层变厚、上皮坏死以及固有层上皮分离；单宁会导致生长中期草鱼肠道杯状细胞增多以及肠道固有层变厚。以上研究结果说明，抗营养因子确实会破坏鱼类肠道结构完整性。在本试验中，我们发现芥酸会导致草鱼肠道充血和肠道绒毛增生，这些病理特征的出现可能跟以下因素有关：①芥酸导致草鱼肠道充血可能跟草鱼肠道中葡萄糖含量升高有关。在大鼠上的研究表明：芥酸会导致脂肪的蓄积，而饲料中高水平的脂肪会增加其血浆中的葡萄糖含量。Sit 等（1980）研究发现，葡萄糖会导致狗肠道充血。因此，我们猜测芥酸可能会导致鱼类肠道脂肪含量增多，进而增加肠道葡萄糖的含量，最终导致鱼类肠道充血。然而，该假设还有待进一步验证。②芥酸会导致草鱼肠道绒毛增生可能跟草鱼肠道中前列腺素 E_2（PGE_2）的增多有关。在大鼠上的研究表明，芥酸会抑制心脏脂肪酸的代谢，这可能会导致大鼠必需脂肪酸的缺乏，而必需脂肪酸缺乏的大鼠肾髓质具有很强的 PGE_2 合成能力。Uribe 和 Johansson（1988）研究发现，高水平的 PGE_2 会导致大鼠肠绒毛增生。因此，我们猜测芥酸可能会促进 PGE2 的合成，进而导致鱼类肠道绒毛增生。然而，该假设也有待进一步验证。总的来说，我们的研究结果表明：芥酸会破坏生长中期草鱼肠道结构完整性。

2.3.4　芥酸增加生长中期草鱼血清中二胺氧化酶活性和 D-乳酸含量

二胺氧化酶是一种能催化组胺、腐胺和尸胺氧化的细胞内酶，其主要存在于肠黏膜中，血液中大部分二胺氧化酶来自肠道。血清中二胺氧化酶活性增加是动物肠道黏膜完整性受损的重要指标。D-乳酸盐是存在于动物肠腔中的细菌代谢产物，完整的肠黏膜具有屏障功能，可防止 D-乳酸渗入门静脉血。因此，血清中 D-乳酸含量的增多也是动物肠道黏膜完整性受损的重要指标。本试验的研究结果表明：草鱼血清中二胺氧化酶活性和 D-乳酸含量随着饲料中芥酸水平的升高而升高。相似的现象在大豆皂甙影响大菱鲆的研究中也有报道，其发现大豆皂甙会增加大菱鲆血浆中二胺氧化酶活性和 D-乳酸含量。由此可见，这进一步证实了芥酸会破坏生长中期草鱼肠道结构完整性。

2.3.5　芥酸对生长中期草鱼肠道细胞结构完整性的影响

研究表明：在外界环境刺激下，动物肠道细胞很容易发生氧化损伤和凋亡，不利于肠道结构完整性的维持。MDA 和 PC 含量的升高以及 DNA 片段化的发生是分别衡量动物肠道细胞氧化损伤和细胞凋亡的重要指标。本试验的研究结果发现：芥酸会导致生长中期草鱼前、中和后肠三个肠段的 MDA 和 PC 含量升高，降低抗氧化酶活力（MnSOD、CAT、GPx、GST 和 GR）以及引起 DNA 片段化的发生，这说明：芥酸会破坏生长中期草鱼肠道细胞结构完整性。有关抗营养因子破坏鱼类肠道细胞结构完整性的现象在我们实验室前期的研究中也有发现，例如，棉酚（$\geqslant 243.94 \text{ mg kg}^{-1}$）和单宁（$\geqslant 30 \text{ g kg}^{-1}$）会导致生长中期草鱼前、中和后肠三个肠段中 MDA 和 PC 含量升高，以及 DNA 片段化的发生。在大豆球蛋白上的研究表明，与对照组相比，8% 的大豆球蛋白会导致幼草鱼中肠和后肠 ROS、MDA 和 PC 明显升高，导致建鲤前、中和后肠脂质过氧化和蛋白质氧化的发生，同时也会导致其三个肠道发生凋亡。在β-伴大豆球蛋白上的研究指出，与对照组相比，8%的β-伴大豆球蛋白会导致建鲤肠道脂质过氧化和蛋白质氧化的发生以及幼草鱼中后肠蛋白质氧化和DNA片段化的发生。除此之外，其他实验室也有类似报道。研究发现，大豆球蛋白（≥30 g·kg^{-1}）和β-伴大豆球蛋白（≥40 g·kg^{-1}）都会使金鲫鱼三个肠段发生脂质氧化损伤。Gu 等（2018）用不同大豆皂甙含量（0、2.5 g·kg^{-1}、7.5 g·kg^{-1}和15 g·kg^{-1}）的饲料饲喂大菱鲆8周，结果发现随着饲料中大豆皂甙含量的升高，鱼体肠道氧化损伤程度增加，并且肠道上皮中发生凋亡的细胞数量也会明显增加。以上研究结果说明：与其他抗营养因子类似，鱼体肠道细胞结构完整性的破坏也是芥酸对鱼类产生负面影响的因素之一。

动物细胞氧化损伤和凋亡分别受 Nrf2 和 p38MAPK（或者 JNK）信号分子的调控。本试验结果表明：芥酸会下调生长中期草鱼前、中和后肠细胞质和细胞核中的 Nrf2 蛋白表达以及上调 *p38MAPK* 的基因表达。进一步的相关性分析结果显示（表 2-7）：本试验中的抗氧化酶相关基因表达（除了 *CuZnSOD* 和 *GSTP2*）与细胞核中的 Nrf2 蛋白表达呈正相关。同时，*caspase-3*（不在后肠）、*caspase-7*、*Apaf-1*、*Bax* 和 *FasL* 的基因表达与 *p38MAPK*（而不是 *JNK*）基因表达呈正相关，以及 *Bcl-2*（不在前肠和后肠）、*IAP* 和 *Mcl-1* 的基因表达与 *p38MAPK*（而不是 *JNK*）的基因表达呈负相关。由此可见：芥酸可能会通过下调细胞核中的 Nrf2 蛋白水平以及上调 *p38MAPK* 的基因表达进而破坏草鱼肠道细胞结构完整性。

表 2-7　生长中期草鱼肠道细胞结构完整性指标相关性分析

自变量	因变量	前肠		中肠		后肠	
		相关系数	P	相关系数	P	相关系数	P
MnSOD mRNA 水平	MnSOD 活性	+0.904	<0.05	+0.893	<0.05	+0.870	<0.05
CAT mRNA 水平	CAT 活性	+0.980	<0.01	+0.964	<0.01	+0.976	<0.01
GPx1a mRNA 水平	GPx 活性	+0.983	<0.01	+0.942	<0.01	+0.905	<0.05
GPx1b mRNA 水平		+0.945	<0.01	+0.934	<0.01	+0.885	<0.05
GPx4a mRNA 水平		+0.957	<0.01	+0.990	<0.01	+0.932	<0.01
GPx4b mRNA 水平		+0.962	<0.01	+0.972	<0.01	+0.968	<0.01
GSTP1 mRNA 水平	GST 活性	+0.981	<0.01	+0.934	<0.01	+0.888	<0.05
GSTO1 mRNA 水平		+0.981	<0.01	+0.983	<0.01	+0.939	<0.01
GSTO2 mRNA 水平		+0.938	<0.01	+0.992	<0.01	+0.960	<0.01

续表

自变量	因变量	前肠 相关系数	P	中肠 相关系数	P	后肠 相关系数	P
GR mRNA 水平	GR 活性	+0.898	<0.05	+0.984	<0.01	+0.937	<0.01
Nuclear Nrf2 蛋白水平	MnSOD mRNA 水平	+0.942	<0.01	+0.937	<0.01	+0.970	<0.01
	CAT mRNA 水平	+0.991	<0.01	+0.988	<0.01	+0.981	<0.01
	GPx1a mRNA 水平	+0.973	<0.01	+0.938	<0.01	+0.958	<0.01
	GPx1b mRNA 水平	+0.938	<0.01	+0.962	<0.01	+0.947	<0.01
	GPx4a mRNA 水平	+0.949	<0.01	+0.922	<0.01	+0.947	<0.01
	GPx4b mRNA 水平	+0.950	<0.01	+0.979	<0.01	+0.967	<0.01
	GSTP1 mRNA 水平	+0.987	<0.01	+0.924	<0.01	+0.966	<0.01
	GSTO1 mRNA 水平	+0.958	<0.01	+0.980	<0.01	+0.971	<0.01
	GSTO2 mRNA 水平	+0.908	<0.05	+0.972	<0.01	+0.935	<0.01
	GR mRNA 水平	+0.973	<0.01	+0.946	<0.01	+0.986	<0.01
Keap1a	Nuclear Nrf2 蛋白水平	−0.977	<0.01	−0.962	<0.01	−0.977	<0.01
P38MAPK	FasL	+0.966	<0.01	+0.966	<0.01	+0.884	<0.05
FasL	caspase-8	+0.996	<0.01	+0.989	<0.01	+0.973	<0.01
caspase-8	caspase-3	+0.969	<0.01	+0.990	<0.01	—	—
	caspase-7	+0.995	<0.01	+0.989	<0.01	+0.986	<0.01
caspase-9	caspase-3	+0.989	<0.01	+0.997	<0.01	—	—
	caspase-7	+0.974	<0.01	+0.994	<0.01	+0.981	<0.01
caspase-2	caspase-3	+0.987	<0.01	+0.996	<0.01	—	—
	caspase-7	+0.996	<0.01	+0.995	<0.01	+0.977	<0.01
Apaf-1	caspase-9	+0.978	<0.01	+0.998	<0.01	+0.980	<0.01
Bax		+0.989	<0.01	+0.995	<0.01	+0.996	<0.01
Bcl-2		—	—	−0.982	<0.01	—	—
Mcl-1		−0.975	<0.01	−0.991	<0.01	−0.959	<0.01
IAP		−0.993	<0.01	−0.996	<0.01	−0.933	<0.01
Apaf-1	caspase-2	+0.995	<0.01	+0.994	<0.01	+0.963	<0.01
Bax		+0.998	<0.01	+0.997	<0.01	+0.984	<0.01
Bcl-2		—	—	−0.992	<0.01	—	—
Mcl-1		−0.993	<0.01	−0.995	<0.01	−0.957	<0.01
IAP		−0.980	<0.01	−0.993	<0.01	−0.953	<0.01

续表

自变量	因变量	前肠		中肠		后肠	
		相关系数	P	相关系数	P	相关系数	P
p38MAPK	Apaf-1	+0.944	<0.01	+0.985	<0.01	+0.967	<0.01
	Bax	+0.969	<0.01	+0.988	<0.01	+0.904	<0.05
	Bcl-2	—	—	-0.988	<0.01	—	—
	Mcl-1	-0.958	<0.01	-0.985	<0.01	-0.924	<0.01
	IAP	-0.989	<0.01	-0.984	<0.01	-0.973	<0.01

关于芥酸对生长中期草鱼肠道氧化损伤方面的影响，我们发现三个有趣的现象。首先，芥酸下调草鱼前、中和后肠三个肠段中的 *MnSOD*（而不是 *CuZnSOD*）的基因表达和酶活力，这可能跟载脂蛋白 A-I 有关。Garner 等（1998）研究发现氧自由基（ROS）的增多会下调人血浆载脂蛋白 A-I 的含量。在小鼠卵巢癌细胞（ID8 细胞）中，低水平的载脂蛋白 A-I 会降低 *MnSOD*（而不是 *CuZnSOD*）的基因表达和酶活力。我们的研究结果表明，芥酸会增加生长中期草鱼三个肠段中的 ROS 含量。因此，我们猜测芥酸可能会增加鱼肠道 ROS 的含量，导致载脂蛋白 A-I 的减少，进而降低 *MnSOD*（而不是 *CuZnSOD*）的基因表达和酶活力。然而，该假设还有待验证。其次，芥酸会降低生长中期草鱼三个肠段中 *GSTP1*（而不是 *GSTP2*）的基因表达，这可能跟细胞外信号调节激酶（ERK）有关。在虹鳟上的研究表明，ROS 水平的升高会上调 *ERK* 的基因表达。Hrubik 等（2016）研究发现 ERK 表达的增多会下调斑马鱼体内 *GSTP1*（而不是 *GSTP2*）的基因表达。我们的研究结果表明芥酸会增加生长中期草鱼三个肠段中的 ROS 含量。因此，我们猜测芥酸可能会导致鱼体肠道 ROS 的增多，进而上调 *ERK* 的基因表达，最终下调 *GSTP1*（而不是 *GSTP2*）的基因表达。然而，该假设也有待进一步验证。最后，芥酸会上调生长中期草鱼三个肠段的 *keap1a*（而不是 *keap1b*）的基因表达，这可能跟脂质过氧化有关。Uchida 等（1998）研究发现脂质过氧化会促进人血清中丙烯醛的产生。在小鼠中，高水平的丙烯醛会通过修饰 Cys-288 激活 keap1。Li 等（2008）研究发现斑马鱼 keap1 包含两个亚型，keap1a 和 keap1b，其含有的半胱氨酸残基分别对应小鼠 keap1 中的 Cys-288 和 Cys-273。我们的研究结果表明芥酸会导致三个肠段脂质过氧化的发生。因此，我们猜测芥酸可能会导致鱼体肠道脂质过氧化的发生，进而上调 *keap1a*（而不是 *keap1b*）的基因表达。然而，该假设也有待进一步验证。同时，与我们实验室之前开展的大豆球蛋白对鱼

类肠道造成氧化损伤的试验结果相比，我们也发现了两个有趣的现象。第一，大豆球蛋白会上调鱼肠道 $MnSOD$ 和 $CuZnSOD$ 的基因表达，而芥酸只下调鱼肠道 $MnSOD$（而不是 $CuZnSOD$）的基因表达，这可能跟二者产生负面影响的靶细胞器不同有关。在菲律宾帘蛤（Ruditapes philippinarum）上的研究表明：MnSOD 位于线粒体基质上，而 CuZnSOD 则发现主要存在于细胞质中。在小鼠上的研究表明，大豆球蛋白会导致过敏症状的发生。Matés 等（1999）研究发现过敏症状的发生会导致人线粒体和细胞质中 SOD 的活性增加。然而，芥酸只会对大鼠的线粒体产生有害影响，这可能只会影响 $MnSOD$（而不是 $CuZnSOD$）的基因表达。因此，我们猜测大豆球蛋白可能会同时对线粒体和细胞质产生有害影响，进而增加鱼肠道中 $MnSOD$ 和 $CuZnSOD$ 的基因表达，而芥酸可能只会对鱼肠道线粒体产生负面影响，进而只下调 $MnSOD$（而不是 $CuZnSOD$）的基因表达。然而，该假设也有待进一步验证。第二，大豆球蛋白会上调鱼肠道 $MnSOD$、$GPX1b$ 和 $GPX4a$ 的基因表达，而芥酸会下调肠道 $MnSOD$、$GPX1b$ 和 $GPX4a$ 的基因表达，这可能跟鱼肠道会对大豆球蛋白造成的刺激产生适应性机制，但对芥酸却没有产生这种机制有关。具体有关产生这种现象的原因，目前还不清楚。我们猜测芥酸对鱼肠道产生的毒性作用可能会比大豆球蛋白产生的毒性作用更强有关。据报道，在海湾扇贝（Argopecten irradians）中，外界刺激会通过增强其体内的抗氧化能力以产生适应性反应，但是，一旦这种平衡被更严重的刺激所打破，抗氧化能力便会降低。这种假设也有待进一步验证。

关于芥酸对生长中期草鱼肠道凋亡方面的影响，我们也发现三个有趣的现象。首先，芥酸上调生长中期草鱼三个肠段 p38MAPK（而不是 JNK）基因表达的影响，这可能跟游离脂肪酸的蓄积有关。Christophersen 和 Bremer（1972）研究发现芥酸会干扰大鼠心脏游离脂肪酸的代谢，这会导致游离脂肪酸的蓄积。在人类方面的研究表明：高水平的游离脂肪酸会增加血管紧张素Ⅱ的含量，后者会激活 p38MAPK（而不是 JNK）信号分子。因此，我们猜测芥酸会通过干扰鱼肠道脂肪酸的代谢，导致游离脂肪酸含量的升高，这会升高血管紧张素Ⅱ的含量，最终促进 p38MAPK（而不是 JNK）的基因表达。然而，该假设还有待进一步验证。其次，芥酸会抑制生长中期草鱼中肠（而不是前肠和后肠）中 $Bcl-2$ 的基因表达，这可能跟油酸在鱼体三个肠段中的被吸收能力不同有关。Thomassen 等（1985）研究发现芥酸在大鼠肠道中会被代谢成油酸。在舌鳞状癌细胞中，油酸会下调 $Bcl-2$ 的基因表达。在大西洋鲑上的研究表明：中肠是吸收油酸的主要位点。因此，我们猜测芥酸在鱼肠道中可能会被代谢成油酸，后者主要在中肠被吸收，进而导致中肠（而不是前肠和后肠）$Bcl-2$ 基因表达的降低。然而，该假设也有待进一步验证。最后，芥酸会上调生长中期草鱼前肠和中肠（而不是后肠）中 $caspase-3$ 的基因表达，这可能跟高脂导致胆囊收缩素的释放有关。Rahman 等（2014）研究发现芥酸会增加大鼠血清中的脂肪水平。脂肪水平的升高会导致泌乳牛胆囊收缩素的释放。在大鼠胰腺腺泡细胞中，Gukovskaya 等（2002）研究发现胆囊收缩素会上调 $caspase-3$ 的基因表达。并且，Feng 等（2012）研究发现胆囊收缩素相关基因主要在草鱼前肠和中肠中表达。因此，我们猜测芥酸可能会导致鱼肠道中脂肪水平的升高，进而促进前肠和中肠（而不是后肠）中胆囊收缩素的释放，最终导致前肠和中肠（而不是后肠）中 $caspase-3$ 基因表达的升高。然而，该假设也有待确定。与我们实验室之前开展的大豆球蛋白对鱼类肠道造成凋亡的试验结果相比，我们也发现了一个有趣的现象。大豆球蛋白会上调鱼体后肠（而不是前肠和中肠）中 $caspase-8$ 和 $caspase-9$ 的基因表达，然而芥酸会上调鱼体前、中和后肠三个肠段中 $caspase-8$ 和 $caspase-9$ 的基因表达，这可能跟大豆球蛋白和芥酸在鱼体三个肠段中的分布和被降解能力不同有关。在体外生化试验中，Vasconcellos 等（2014）研究发现大豆球蛋白降解成的小肽具有抗氧化特性。在仔猪上的研究表明，大豆球蛋白主要在近端消化道中被降解。因此，我们猜测大豆球蛋白上调鱼后肠（而不是前肠和中肠）中 $caspase-8$ 和 $caspase-9$ 的基因表达可能跟鱼肠道三个肠段对大豆球蛋白的降解能力不同有关。而我们的研究发现芥酸会均匀地蓄积在生长中期草鱼三个肠段中，这会导致鱼三个肠段中 $caspase-8$ 和 $caspase-9$ 的基因表达升高。然而，该假设也有待验证。

2.3.6 芥酸对生长中期草鱼肠道细胞间结构完整性的影响

紧密连接蛋白和黏附连接蛋白是动物肠道细胞间结构的重要组成成分，二者相互偶联，互相影响，其中，紧密连接蛋白主要包括环状蛋白 ZO-1 和跨膜蛋白 occludin 等，黏附连接蛋白主要包括 E-cadherin 和 nectin 等。已有的研究结果表明：动物肠道细胞间结构完整性的破坏与紧密连接和黏附连接相关蛋白表达下降有关。同时，离子通道蛋白 claudin-15 表达的升高会增加小鼠肠道细胞间的间隙，不利于动物肠道细胞间结构完整性的形成。本试验的研究结果发现：芥酸会下调生长中期草鱼前、中和后肠紧密连接蛋白 $ZO-1$、$ZO-2$、$occludin$、$claudin-b$、$claudin-c$、$claudin-f$、$claudin-3c$、$claudin-7a$、$claudin-7b$ 和 $claudin-11$ 的基因表达，以及黏附连接蛋白 $\alpha-catenin$、$\beta-catenin$、$E-cadherin$、$nectin$ 和 $afadin$ 的基因表达，和上调离子通道蛋白 $claudin-15a$ 的基因表达。同时，我们发现芥酸会降低生长中期草鱼前、中和后肠中紧密连接蛋白 ZO-1 以及黏附连接蛋白 E-cadherin 和 β-catenin 的蛋白表达。这些研究结果表明：芥酸会破坏生长中期草鱼肠道细胞间结构完整性。抗营养因子破坏鱼体肠道细胞间结构完整性的现象在其他试验中也有报道，例如，棉酚和单宁都会破坏生长中期草鱼肠道细胞间结构完整性。在建鲤上的研究表明：大豆球蛋白会下调其中肠 *occludin* 和 *claudin-3c* 的基因表达。Gu 等（2018）用大豆皂甙含量为 0、2.5 g/kg、7.5 g/kg 和 15 g/kg 的饲料饲喂大菱鲆 3 周，通过透射电镜电镜发现大豆皂甙会导致草鱼肠道细胞间紧密连接蛋白和黏附连接蛋白变短，并且紧密连接蛋白和黏附连接蛋白存在发育不完整的现象。

动物肠道细胞间紧密连接蛋白和黏附连接蛋白主要受 MLCK、NMII 和 ROCE 信号分子的影响，后三者又都受 RhoA 信号分子的调控。本试验研究结果表明：芥酸会上调生长中期草鱼前、中和后肠 *MLCK*、*NMII* 和 *ROCK* 的基因表达，同时会增加三个肠段的 RhoA 活性。进一步的相关性分析结果显示（表 2-8）：*ROCK*、*NMII* 和 *MLCK* 的基因表达与草鱼肠细胞间紧密连接蛋白（离子通道蛋白 *claudin-15a* 呈正相关）和黏附连接蛋白的基因表达呈负相关，并且 RhoA 活性与 *ROCK*、*NMII* 和 *MLCK* 的基因表达呈正相关。由此可见：芥酸可能会激活 RhoA 信号分子上调 *MLCK*、*NMII* 和 *ROCK* 的基因表达，从而破坏草鱼肠道细胞间结构完整性。然而，该假设还有待进一步验证。

有趣的是，芥酸不影响生长中期草鱼前、中和后肠离子通道蛋白 *claudin-15b* 和 *claudin-12* 的基因表达。首先，芥酸只上调生长中期草鱼肠道三个肠段 *claudin-15a*（而不是 *claudin-15b*）的基因表达，这可能跟芥酸导致草鱼肠道葡萄糖含量增多有关。Tamura 等（2011）研究发现小鼠肠道中 claudin-15 形成的细胞旁路通道主要负责葡萄糖的吸收。Vemuri 等（2018）研究发现芥酸会导致雄性叙利亚金黄地鼠血浆中葡萄糖的含量升高。Bossus 等（2015）研究发现在日本青鳉（*Oryzias latipes*）肠道中，*claudin-15a* 的 mRNA 水平明显高于 *claudin-15b*。因此，我们猜测芥酸只上调生长中期草鱼肠道三个肠段 *claudin-15a*（而不是 *claudin-15b*）的基因表达，这可能跟其适应由芥酸导致鱼类肠道葡萄糖含量升高有关。然而，该假设还有待进一步验证。其次，芥酸不影响生长中期草鱼肠道 *claudin-12* 的基因表达，这可能跟芥酸不影响草鱼肠道细胞内钙离子浓度有关。Fujita 等（2008）研究发现在人肠道上皮细胞中，claudin-12 负责钙离子的吸收。Heiskanen 和 Savolainen（1997）研究发现芥酸不影响人多形核白细胞内的钙离子水平。因此，我们猜测芥酸可能不影响草鱼肠道细胞内的钙离子水平，从而不影响其 *claudin-12* 的基因表达。然而，该假设也有待进一步验证。与我们实验室之前开展的大豆球蛋白对鱼类肠道造成细胞间结构完整性破坏的试验结果相比，我们也发现了两个有趣的现象。首先，大豆球蛋白会上调鱼肠道中 $claudin-11$ 的基因表达，而芥酸会下调鱼肠道中 $claudin-11$ 的基因表达，这可能跟皮质醇有关。Jiang 等（2015）研究发现大豆球蛋白会上调建鲤肠道中 $IL-1\beta$ 的基因表达。在豚鼠上，$IL-1\beta$ 含量的增多会导致皮质醇水平的升高。在河豚鱼上的研究表明：皮质醇会上调鱼体鳃上皮细胞中 $claudin-11a$ 的基因表达。然而，芥酸会增加大鼠游离脂肪酸的含量。游离脂肪酸的增多会降低人皮质醇的水平。因此，我们猜测大豆球蛋白可能会增加鱼肠道中的皮质醇含量，进而上调 $claudin-11$ 的基因表达，而芥酸可能会降低鱼肠道中皮质醇的含量，进而下调 $claudin-11$ 的基因表达。然而，该假设也有待进一步验证。其次，大豆球蛋白对鱼体前肠和中肠（而不是后肠）中 $claudin-7$ 的基因表达没有影响，但是芥酸会下调鱼体三个肠段中 $claudin-7$ 的基因表达。这两种抗营养因子对鱼体肠段 $claudin-7$ 的基因表达存在的差异，目前还不清楚，有待研究。

表 2-8　生长中期草鱼肠道细胞间结构完整性指标相关性分析

自变量	因变量	前肠		中肠		后肠	
		相关系数	P	相关系数	P	相关系数	P
ROCK	$ZO-1$	-0.929	<0.01	-0.999	<0.01	-0.981	<0.01
	$ZO-2$	-0.986	<0.01	-0.970	<0.01	-0.973	<0.01
	$occludin$	-0.945	<0.01	-0.979	<0.01	-0.986	<0.01
	$claudin-b$	-0.882	<0.05	-0.979	<0.01	-0.984	<0.01
	$claudin-c$	-0.924	<0.01	-0.990	<0.01	-0.983	<0.01
	$claudin-f$	-0.959	<0.01	-0.989	<0.01	-0.979	<0.01
	$claudin-3c$	-0.926	<0.01	-0.986	<0.01	-0.979	<0.01
	$claudin-7a$	-0.943	<0.01	-0.992	<0.01	-0.986	<0.01
	$claudin-7b$	-0.840	<0.05	-0.935	<0.01	-0.953	<0.01
	$claudin-11$	-0.938	<0.01	-0.998	<0.01	-0.989	<0.01
	$claudin-15a$	+0.949	<0.01	+0.983	<0.01	+0.976	<0.01
	$\alpha-catenin$	-0.972	<0.01	-0.966	<0.01	-0.957	<0.01
	$\beta-catenin$	-0.937	<0.01	-0.992	<0.01	-0.974	<0.01
	$E-cadherin$	-0.979	<0.01	-0.995	<0.01	-0.981	<0.01
	$nectin$	-0.980	<0.01	-0.980	<0.01	-0.989	<0.01
ROCK	$afadin$	-0.984	<0.01	-0.990	<0.01	-0.976	<0.01

续表

自变量	因变量	前肠		中肠		后肠	
		相关系数	P	相关系数	P	相关系数	P
NMII	ZO-1	−0.957	<0.01	−0.943	<0.01	−0.989	<0.01
	ZO-2	−0.996	<0.01	−0.876	<0.05	−0.983	<0.01
	occludin	−0.968	<0.01	−0.869	<0.05	−0.986	<0.01
	claudin-b	−0.911	<0.05	−0.872	<0.05	−0.998	<0.01
	claudin-c	−0.955	<0.01	−0.910	<0.05	−0.990	<0.01
	claudin-f	−0.983	<0.01	−0.972	<0.01	−0.982	<0.01
	claudin-3c	−0.948	<0.01	−0.924	<0.01	−0.977	<0.01
	claudin-7a	−0.971	<0.01	−0.908	<0.05	−0.998	<0.01
	claudin-7b	−0.863	<0.05	−0.820	<0.05	−0.931	<0.01
	claudin-11	−0.962	<0.01	−0.930	<0.01	−0.991	<0.01
	claudin-15a	+0.965	<0.01	+0.893	<0.05	+0.977	<0.01
	α-catenin	−0.973	<0.01	−0.840	<0.05	−0.947	<0.01
	β-catenin	−0.960	<0.01	−0.916	<0.05	−0.983	<0.01
	E-cadherin	−0.977	<0.01	−0.926	<0.01	−0.970	<0.01
	nectin	−0.994	<0.01	−0.886	<0.05	−0.991	<0.01
	afadin	−0.996	<0.01	−0.896	<0.05	−0.979	<0.01
MLCK	ZO-1	−0.988	<0.01	−0.972	<0.01	−0.959	<0.01
	ZO-2	−0.978	<0.01	−0.961	<0.01	−0.964	<0.01
	occludin	−0.993	<0.01	−0.976	<0.01	−0.991	<0.01
	claudin-b	−0.981	<0.01	−0.995	<0.01	−0.946	<0.01
	claudin-c	−0.986	<0.01	−0.989	<0.01	−0.984	<0.01
	claudin-f	−0.974	<0.01	−0.944	<0.01	−0.989	<0.01
	claudin-3c	−0.987	<0.01	−0.954	<0.01	−0.991	<0.01
	claudin-7a	−0.955	<0.01	−0.974	<0.01	−0.963	<0.01
	claudin-7b	−0.921	<0.01	−0.982	<0.01	−0.982	<0.01
	claudin-11	−0.990	<0.01	−0.980	<0.01	−0.988	<0.01
	claudin-15a	+0.997	<0.01	+0.978	<0.01	+0.989	<0.01
	α-catenin	−0.961	<0.01	−0.968	<0.01	−0.921	<0.01
	β-catenin	−0.973	<0.01	−0.961	<0.01	−0.977	<0.01

续表

自变量	因变量	前肠		中肠		后肠	
		相关系数	P	相关系数	P	相关系数	P
MLCK	E-cadherin	−0.959	<0.01	−0.971	<0.01	−0.979	<0.01
	nectin	−0.967	<0.01	−0.958	<0.01	−0.973	<0.01
	afadin	−0.970	<0.01	−0.964	<0.01	−0.973	<0.01
GTP-RhoA/Total RhoA	ROCK	+0.961	<0.01	+0.984	<0.01	+0.985	<0.01
	NMII	+0.981	<0.01	+0.957	<0.01	+0.995	<0.01
	MLCK	+0.934	<0.01	+0.911	<0.05	+0.973	<0.01

2.3.7 生长中期草鱼饲料中芥酸的控制剂量

本研究结果表明：芥酸会降低生长中期草鱼的生产性能。如图2-14所示，以增重百分比、前肠 MDA 含量、中肠 ROS 含量以及后肠 PC 含量为标识，采用折线回归分析，确定生长中期草鱼（129.17~471.18 g）饲料中芥酸的控制剂量分别为 0.64%、0.48%、0.48% 和 0.53%。

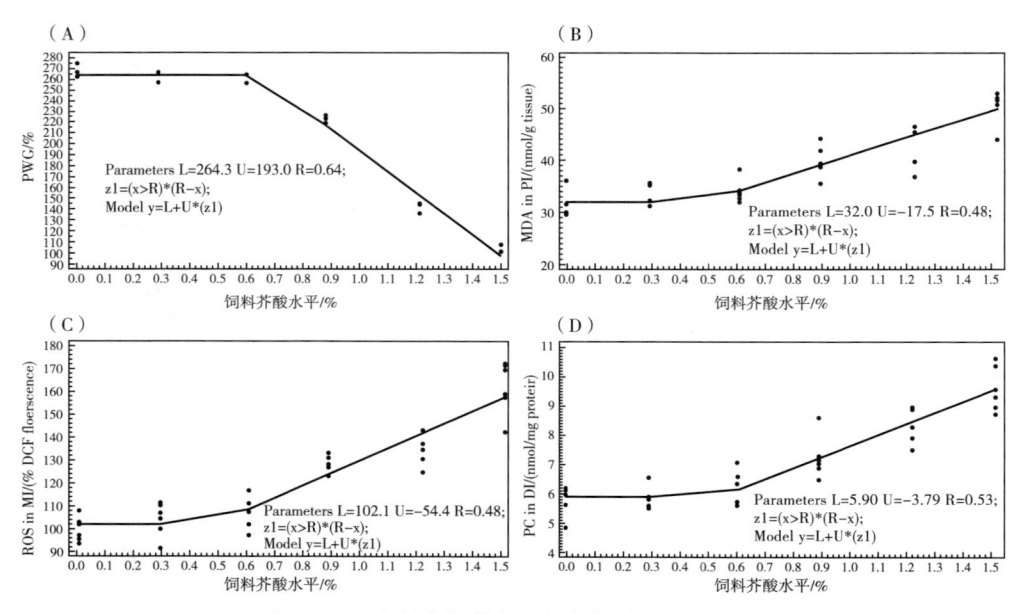

图2-14 生长中期草鱼饲料中芥酸的控制剂量

参数定义如下：L 为平台；U 为坡度；R 为断点

2.4 小结

根据试验结果得出如下结论：

（1）芥酸会降低生长中期草鱼的生长性能和饲料利用效率，根据增重百分比、前肠 MDA 含量、中肠 ROS 含量以及后肠 PC 含量，确定生长中期草鱼（129.17～471.18 g）饲料中芥酸的控制剂量分别为 0.64%、0.48%、0.48% 和 0.53%。

（2）芥酸会引起生长中期草鱼肠道氧化损伤，这可能与芥酸抑制草鱼肠道 Nrf2/keap1a（不是 keap1b）信号通路，下调 *MnSOD*（不是 *CuZnSOD*）、*CAT*、*GPx1a*、*GPx1b*、*GPx4a*、*GPx4b*、*GSTP1*（不是 *GSTP2*）和 *GSTO1* 基因表达，降低 MnSOD（不是 CuZnSOD）、CAT、GPx、GR、GST、ASH 和 AHR 的活性和 GSH 的含量，导致脂质过氧化（MDA 增多）和蛋白质氧化（PC 增多）有关。

（3）芥酸会引起生长中期草鱼肠道细胞凋亡，这可能与芥酸激活 p38MAPK（不是 JNK）信号途径，上调 *FasL*、*Apaf-1*、*Bax*、*caspase-2*、*caspase-3*（不在后肠）、*caspase-7*、*caspase-8* 和 *caspase-9* 的基因表达，下调 *IAP*、*Mcl-1* 和 *Bcl-2*（不在前肠和后肠）的基因表达有关。

（4）芥酸会破坏生长中期草鱼肠道细胞间结构完整性，这可能与芥酸激活 RhoA 信号途径，上调关键信号分子 *MLCK*、*ROCK* 和 *NMII* 的基因表达，下调紧密连接（*ZO-1*、*ZO-2*、*occludin*、*claudin-b*、*claudin-c*、*claudin-f*、*claudin-3c*、*claudin-7a*、*claudin-7b* 和 *claudin-11* 和黏附连接（α-catenin、β-catenin、*E-cadherin*、*nectin* 和 *afadin*）的基因表达，上调离子通道蛋白 *claudin-15a*（不是 *claudin-15b* 和 *claudin-12*）的基因表达，以及下调 E-cadherin、β-catenin 和 ZO-1 的蛋白表达有关。

第三章 芥酸激活 RhoA 信号途径

　　第二章中的研究结果表明，芥酸会破坏生长中期草鱼肠道细胞间结构的完整性以及细胞结构的完整性。在水产动物中，尽管已有大量关于抗营养因子对肠道细胞结构的研究，但针对抗营养因子如何影响动物肠道细胞间结构的研究仍较为有限。肠道细胞间结构对动物的生长与健康至关重要。它不仅能有效阻止肠腔中的病原菌、毒素和抗原物质的侵害，还在调节药物分子、水、营养物质和离子摄入方面发挥着重要作用，从而维持肠道的稳态。虽然我们实验室此前的研究已经表明，抗营养因子如棉酚和单宁会破坏生长中期草鱼肠道细胞间结构的完整性，但这些现象的具体分子机制仍不完全明确，需要进一步探索。动物肠道细胞间的结构主要由紧密连接蛋白和黏附连接蛋白组成。紧密连接由 40 多种蛋白质构成，其中包括跨膜蛋白（如 occludin 和 claudin）和外周膜蛋白（如 ZO 蛋白）。跨膜蛋白负责细胞与细胞之间的接触，而外周膜蛋白则与肌动蛋白细胞骨架连接。黏附连接蛋白主要由 nectins 和 cadherins 两大跨膜蛋白家族组成，其细胞外区域介导细胞与相邻细胞的黏附，而细胞内区域则与一系列其他蛋白质相互作用。紧密连接的组装和黏附连接的形成是相互耦合的，二者相互影响。RhoA 信号通路在调控动物肠道细胞间紧密连接蛋白和黏附连接蛋白中发挥着至关重要的作用。研究表明，RhoA 活性的增加会抑制小鼠足细胞中紧密连接蛋白 ZO-1 的表达，大鼠脑微血管内皮细胞中 ZO-1 和 occludin 的表达，以及人肾小球内皮细胞中 occludin 的表达。同样的现象也出现在人结肠癌细胞的研究中，表明紧密连接蛋白 ZO-1、occludin 和黏附连接蛋白 E-cadherin 的蛋白表达降低与 RhoA 活性的升高密切相关。在第二章中，我们也观察到芥酸会显著增加生长中期草鱼三个肠段中的 RhoA 活性。然而，芥酸破坏草鱼肠道细胞间结构完整性是否与 RhoA 信号通路的激活直接相关，尚不明确。

　　因此，我们计划在后续研究中开展体外实验，通过阻断 RhoA 信号分子，进一步验证芥酸是否通过激活 RhoA 信号通路而破坏草鱼肠道细胞间结构的完整性。这一研究有助于揭示芥酸影响草鱼肠道健康的分子机制，为提高水产养殖中菜籽粕的利用效率提供理论支持。

3.1 材料与方法

3.1.1 试验设计

本部分包括 2 个细胞试验。

（1）考察不同芥酸浓度对草鱼肠细胞间紧密连接蛋白和黏附连接蛋白相关基因表达、细胞培养液中乳酸脱氢酶活性和细胞中 MDA 含量的影响，以确定能引起草鱼肠细胞间结构完整性破坏的芥酸浓度。试验共设计 6 个芥酸浓度梯度（0、1 mmol/L、2 mmol/L、3 mmol/L、4 mmol/L 和 5 mmol/L），每个浓度 6 个重复，具体试验设计见表 3-1。

表 3-1 试验设计 1

处理	1	2	3	4	5	6
重复数	6	6	6	6	6	6
EA/（mmol·L^{-1}）	0	1	2	3	4	5

（2）通过添加 RhoA 活性抑制剂，考察芥酸对草鱼肠细胞间结构完整性的影响及机制，试验共设计 4 个处理，每个处理 6 个重复，分别为对照组（Control）、RhoA 活性抑制剂组（Rhosin hydrochloride+）、芥酸组（EA+）和芥酸+RhoA 活性抑制剂组（EA+ Rhosin hydrochloride+），具体试验设计见表 3-2。

表 3-2 试验设计 2

处理	对照组	对照组+抑制剂	处理组	处理组+抑制剂
重复数	6	6	6	6
DMSO	+	−	+	−
RhoA 活性抑制剂	−	+	−	+
EA	−	−	+	+

3.1.2 试验材料

3.1.2.1 仪器与设备

紫外可见分光光度计 UV-1100（上海美谱达仪器有限公司，中国）、尼康相差倒置显微镜（Nikon，日本）、细胞培养板（Corning，美国）和生物安全柜（苏州安泰，中国）等，其他仪器和设备同第二章。

3.1.2.2 试剂与药品

芥酸（sigma，美国）、无脂肪酸牛血清白蛋白（sigma，美国）、二甲基亚砜（sigma，美国）、鼠尾胶原蛋白 I 型（solarbio，美国）、胶原蛋白酶（sigma，美国）、中性蛋白酶（sigma，美国）、氯化钾（上海国药，中国）、磷酸二氢钾（上海国药，中国）、六水氯化镁（上海国药，中国）、七水硫酸镁（上海国药，中国）、氯化钠（上海国药，中国）、胎牛血清（Gibco，美国）、M199（武汉普诺赛，中国）和 RhoA 活性抑制剂（Rhosin hydrochloride）（上海睿铂赛，中国）。其他试剂和药品同第二章。

3.1.2.3 草鱼肠细胞培养

草鱼肠道细胞的分离主要参照段绪东（2019）的方法进行。主要操作过程如下：①选取体重 50 g 左右的健康草鱼，用乙醇擦拭鱼体表消毒；②用解剖剪破坏草鱼脊髓，剖开腹腔，将肠道取出，并剔除肠道上的黏附物；③用含有三抗的 HANKs 液反复冲洗草鱼肠腔，用眼科剪剖开肠道，并将其剪碎成 1 mm^3 的组织小块；④将这些组织小块再用含三抗的 HANKs 液反复冲洗，然后用胶原酶Ⅳ（30 μg/mL）和中性蛋白酶Ⅱ（20 μg/mL）联合消化 70 min；⑤用 S–DMEM（含 2%山梨醇的 DMEM）提取细胞团，然后离心，倒掉上清，将离心管底部的细胞团稀释于完全培养基中，细胞接种于胶原蛋白包被的培养板中并在培养箱（28 ℃）中培养。

3.1.3 芥酸配制

称取一定量的芥酸溶于无水乙醇中，并采用 70℃水浴促进芥酸溶解，配制浓度为 30 mol/L 的芥酸储备液。另外配制 10%不含脂肪酸的牛血清蛋白溶液，滤灭，用于将 30 mol/L 的芥酸储备液稀释成浓度为 30 mmol/L 的工作液并充分振荡混匀，−20℃保存。

3.1.4 RhoA 抑制剂配制

Rhosin hydrochloride 溶于 DMSO 中，得到浓度为 30 mmol/L 的储备液，

−20℃条件下备用。使用前，按1000倍稀释。

3.1.5 细胞处理

①细胞接种于24孔板中培养48 h后，添加含不同浓度的芥酸（1 mmol/L、2 mmol/L、3 mmol/L、4 mmol/L和5 mmol/L），阴性对照为不加芥酸组（0 mmol/L），每一浓度设置6个重复孔，处理肠细胞24 h，随后收集各芥酸浓度处理的细胞，筛选出能破坏草鱼肠道细胞间结构完整性的芥酸浓度；②将草鱼肠细胞接种于12孔板中培养48 h后，先添加30 μmol/L抑制剂Rhosin hydrochloride（预试验确定浓度）处理2 h，然后再添加3 mmol/L芥酸（由①中筛选出的浓度）处理24 h，阴性对照为不添加芥酸组（0 mmol/L）。

3.1.6 观测指标

3.1.6.1 乳酸脱氢酶（LDH）活性检测

芥酸应激处理24 h后，收集草鱼肠细胞培养液，用于LDH活性检测。参照Mulier等（1998）的方法，在丙酮酸钠形成乳酸钠的过程中，根据还原烟酰胺腺嘌呤二核苷酸（NADH）在340 nm下吸光度的减少率来确定LDH的活性。

3.1.6.2 丙二醛（MDA）含量检测

芥酸应激处理24 h后，收集草鱼肠细胞，用于MDA含量检测。参照甘雷（2016）的方法，草鱼肠细胞中MDA能与硫代巴比妥酸（TBA）反应生成红色物质，后者在532 nm下吸光度的增加量来确定MDA的含量。

3.1.6.3 基因表达

芥酸处理24 h后，弃掉细胞培养液，添加200 μL/孔的RNAiso Plus试剂对细胞进行裂解，收集细胞裂解液。RNA提取、cDNA合成以及荧光定量的操作同第二章，引物序列同第二章。

3.1.6.4 蛋白表达

同第二章。

3.1.6.5 RhoA 活性检测

同第二章。

3.1.7 统计分析

本试验所有数据用平均值±标准误（Mean±SE）表示。芥酸对草鱼肠细胞中 MDA 含量和培养液中 LDH 活性作用采用 SAS 软件（SAS Institute，Inc.，2006）的单因素方差分析，并结合 DUNCAN 法进行多重比较。对芥酸与 RhoA 活性抑制剂的作用采用 SAS 软件的 2×2 多因素进行分析，并对处理组间的差异进行方差分析。$P \leqslant 0.05$ 为差异显著，$0.05<P \leqslant 0.1$ 为差异显著趋势。

3.2 试验结果

不同芥酸浓度对草鱼肠道细胞间结构完整性的影响见表 3-3 和图 3-1。当芥酸浓度达到 3 mmol/L 时，会显著提高草鱼肠细胞中 MDA 的含量以及细胞培养液中 LDH 的活性（$P<0.05$）。同时，2 mmol/L 浓度的芥酸会显著下调草鱼肠细胞间 ZO-1、occludin、claudin-b、JAM、α-catenin 和 β-catenin 的 mRNA 水平以及上调 claudin-15a、NMII 和 ROCK 的 mRNA 水平（$P<0.05$）。3 mmol/L 浓度的芥酸会显著下调草鱼肠细胞间 E-cadherin 的 mRNA 水平（$P<0.05$）。

表 3-3 不同芥酸浓度对草鱼肠细胞中 MDA 含量和细胞培养液中 LDH 活性的影响

	芥酸含量/（mmol·L^{-1}）					
	0	1	2	3	4	5
MDA	0.25±0.04^a	0.30±0.02^{ab}	0.30±0.04^b	0.50±0.06^c	0.52±0.06^c	0.65±0.04^d
LDH	47.25±5.04^a	48.07±7.06^a	51.09±5.06^a	60.64±5.95^b	70.57±4.47^c	79.25±4.16^d

注 数据采用平均值±标准误表示（$n=6$），不同字母在同一行表示差异显著（$P<0.05$）。MDA（nmol/mg 蛋白）；LDH（U/mg 蛋白）。

芥酸对草鱼肠细胞中 RhoA 活性的影响如图 3-2 所示。芥酸显著增加草鱼肠细胞中 GTP-RhoA/Total-RhoA 的比值，而 RhoA 活性抑制剂（Rhosin hydrochloride）可显著下调由芥酸上调的 GTP-RhoA/Total-RhoA 的比值（$P<0.05$）。

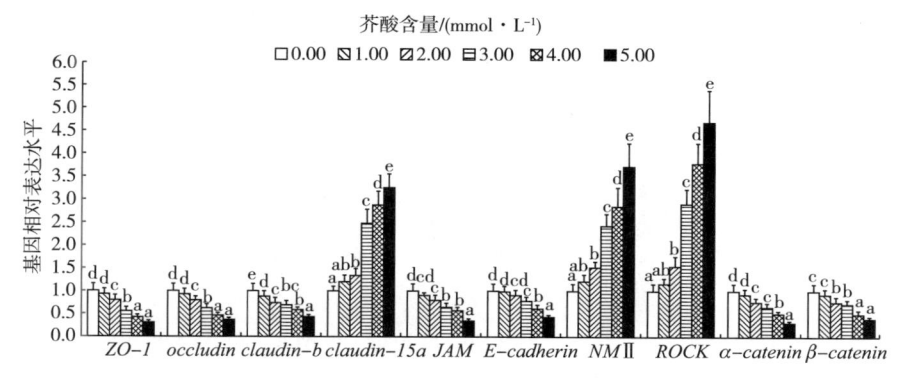

图 3-1　不同芥酸浓度对草鱼肠细胞间紧密连接和黏附连接相关基因 mRNA 水平的影响

数据表示为平均值±标准误（$n=6$）；图柱上的不同字母表示显著差异（$P<0.05$）

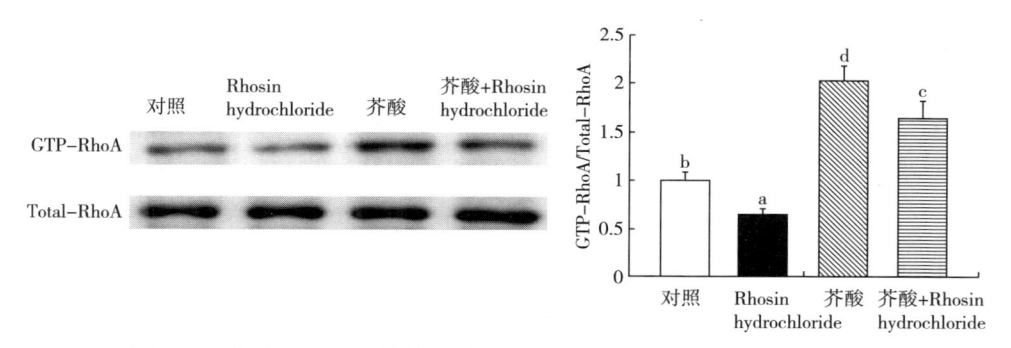

图 3-2　芥酸和 RhoA 活性抑制剂（Rhosin hydrochloride）处理草鱼肠细胞后对草鱼肠细胞 GTP-RhoA/Total-RhoA 比值的影响

数据表示为平均值±标准误（$n=6$）；图柱上的不同字母表示显著差异（$P<0.05$）

芥酸对草鱼肠细胞间紧密连接蛋白和黏附连接蛋白相关基因 mRNA 水平的影响如图 3-3 所示。与对照组相比，芥酸下调草鱼肠细胞间紧密连接蛋白 ZO-1、occludin、claudin-7a 和 claudin-7b（$P<0.05$），以及黏附连接蛋白 nectin、E-cadherin 和 β-catenin 的 mRNA 水平（$P<0.05$），同时上调离子通道蛋白 claudin-15a 以及关键信号分子 MLCK、NMII 和 ROCK 的 mRNA 水平（$P<0.05$），而 RhoA 活性抑制剂（Rhosin hydrochloride）可显著上调由芥酸下调的 ZO-1、occludin、claudin-7a、claudin-7b、nectin 和 E-cadherin 的 mRNA 水平（$P<0.05$），同时下调由芥酸上调的 claudin-15a、MLCK、NMII 和 ROCK 的 mRNA 水平（$P<0.05$）。与此同时，RhoA 活性抑制剂（Rhosin hydrochloride）有上调由芥酸下调的 β-catenin 的 mRNA 水平的趋势（$P=0.072$）。

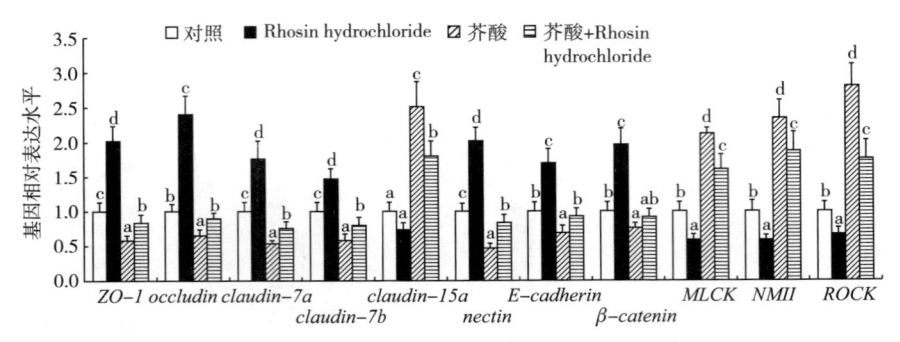

图 3-3 芥酸和 RhoA 活性抑制剂（Rhosin hydrochloride）处理草鱼肠细胞后对草鱼肠细胞间紧密连接、黏附连接和关键信号分子相关基因 mRNA 水平的影响

数据表示为平均值±标准误（$n=6$）；图柱上的不同字母表示显著差异（$P<0.05$）

芥酸对草鱼肠细胞间 E-cadherin、β-catenin 和 ZO-1 蛋白表达的影响如图 3-4 所示。与对照组相比，芥酸可下调草鱼肠细胞间 E-cadherin、β-catenin 和 ZO-1 的蛋白表达（$P<0.05$），而 RhoA 活性抑制剂（Rhosin hydrochloride）可显著上调由芥酸下调的 E-cadherin、β-catenin 和 ZO-1 的蛋白表达（$P<0.05$）。

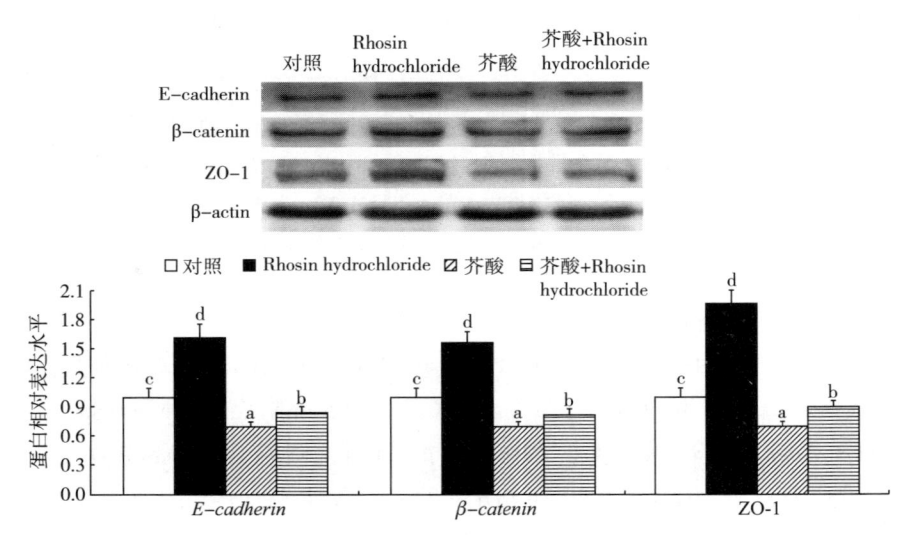

图 3-4 芥酸和 RhoA 活性抑制剂（Rhosin hydrochloride）处理草鱼肠细胞后对草鱼肠细胞间 E-cadherin、β-catenin 和 ZO-1 蛋白表达的影响

数据表示为平均值±标准误（$n=6$）；图柱上的不同字母表示显著差异（$P<0.05$）

3.3 讨论

为进一步研究芥酸破坏草鱼肠道细胞间结构完整性的机制，我们通过体外试验，构建了芥酸破坏草鱼肠细胞间结构完整性模型。首先采用不同浓度的芥酸（0、1 mmol/L、2 mmol/L、3 mmol/L、4 mmol/L 和 5 mmol/L）对草鱼肠细胞处理 24 h，筛选出能破坏草鱼肠细胞间结构完整性的芥酸浓度。LDH 是反映动物细胞结构完整性的关键酶，其活力增加，说明结构损伤严重。MDA 是细胞氧化损伤的关键指标，MDA 含量增多，说明氧化损伤程度加重。本试验的结果表明：当芥酸浓度增加到 3 mmol/L 时，草鱼肠细胞培养液中 LDH 活性以及肠细胞中 MDA 含量明显提高，同时草鱼肠细胞间紧密连接蛋白 *ZO-1*、*occludin*、*claudin-b* 和 *JAM*，以及黏附连接蛋白 *E-cadherin*、*α-catenin* 和 *β-catenin* 的基因表达都明显降低，这说明：添加 3 mmol/L 的芥酸可以成功构建草鱼肠细胞间结构完整性破坏模型。因此，我们利用 3 mmol/L 浓度的芥酸和 RhoA 活性抑制剂 Rhosin hydrochloride 研究芥酸破坏草鱼肠细胞间结构完整性的机制。

在本试验中，芥酸会显著下调草鱼肠细胞间 *ZO-1*、*occludin*、*claudin-7a*、*claudin-7b*、*nectin*、*E-cadherin* 和 *β-catenin* 的基因表达，上调 *claudin-15a* 的基因表达，和下调 ZO-1、E-cadherin 和 β-catenin 的蛋白表达，同时显著增加了 RhoA 活性。这说明，芥酸可能通过激活 RhoA 信号途径破坏草鱼肠细胞间结构完整性。Rhosin hydrochloride 是 RhoA 的活性抑制剂，其被广泛应用于 RhoA 生物学功能研究中。在本试验中，与单独添加芥酸相比，在芥酸破坏草鱼肠细胞间结构完整性前提前添加 RhoA 活性抑制剂 Rhosin hydrochloride 可显著上调草鱼肠细胞间 *ZO-1*、*occludin*、*claudin-7a*、*claudin-7b*、*E-cadherin* 和 *β-catenin* 的基因表达，下调 *claudin-15b* 的基因表达，以及上调 ZO-1、E-cadherin 和 β-catenin 的蛋白表达。这说明：芥酸可通过 RhoA 信号途径破坏草鱼肠细胞间结构完整性。动物肠道细胞间结构完整性受 RhoA 信号分子调控的现象在其他试验中也有发现。例如，RhoA 信号分子的激活会下调人结肠癌细胞间 E-cadherin 和 α-catenin 的蛋白表达，下调小鼠结肠细胞间紧密连接蛋白 ZO-1 和 occludin 的蛋白表达以及大鼠结肠细胞间紧密连接 occludin 和 claudin 的表达分布。综上所述，本试验进一步通过体外试验验证了芥酸可通过 RhoA 信号途径，下调 *ZO-1*、*occludin*、*claudin-7a*、*claudin-7b*、*E-cadherin*、*β-catenin* 和 *nectin* 的基因表达，上调离子通道蛋白 *claudin-15b* 的基因表达，和下调 ZO-1、E-cadherin 和 β-catenin 的蛋白表达，进而破坏草鱼肠细胞间结构完整性。

3.4 小结

根据试验结果，得出以下结论：

（1）通体外试验证明，芥酸破坏了草鱼肠细胞间的结构完整性。

（2）芥酸破坏草鱼肠细胞间结构完整性与芥酸激活 RhoA 信号通路，进一步上调关键信号分子 *MLCK*、*ROCK* 和 *NMII* 的基因表达，下调紧密连接蛋白（*ZO-1*、*occludin*、*claudin-7a* 和 *claudin-7b*）和黏附连接蛋白（*E-cadherin*、*β-catenin* 和 *nectin*）的基因表达，上调离子通道蛋白 *claudin-15a* 的基因表达，以及下调 E-cadherin、β-catenin 和 ZO-1 的蛋白表达有关。

第四章 芥酸对内质网应激的影响

第三章的研究结果表明，芥酸能够激活 RhoA 信号通路，从而破坏草鱼肠道细胞间结构的完整性。动物肠道细胞间结构是维持肠道屏障功能的重要部分。已有研究表明，内质网应激的发生会导致动物肠道物理屏障功能受损。例如，在小鼠肠道中，内质网应激能够降低紧密连接蛋白 ZO-1 和 occludin 的表达，这提示内质网应激可能在芥酸破坏草鱼肠道细胞间结构完整性中发挥重要作用。内质网应激本质上是动物对外界刺激的一种适应性反应，旨在清除由外部应激因素引起的内质网腔内非折叠蛋白和错误折叠蛋白的积累，从而减轻外界压力对细胞的负面影响。然而，当这种应激反应未能及时解决时，内质网应激会对动物产生一系列不良影响，包括脂肪沉积、细胞凋亡以及肠道屏障功能受损等。尽管如此，关于内质网应激是否参与芥酸破坏草鱼肠道细胞间结构的机制，目前尚未有明确的研究报道。因此，我们计划首先通过同一生长实验考察芥酸对生长中期草鱼肠道细胞超微结构和内质网应激相关指标的影响。根据实验结果，我们将进一步开展细胞实验，以验证内质网应激是否在芥酸破坏草鱼肠道细胞间结构完整性中发挥作用。

内质网应激的主要信号通路包括：PERK/eIF2α、IRE1/XBP1 和 ATF6，而非折叠蛋白反应（UPR）则是内质网应激的下游信号之一。已有研究表明，内质网应激会显著降低大鼠血脊髓屏障中紧密连接蛋白 occludin 和 β-catenin 的表达，并减少人气管上皮细胞中 ZO-1 和黏附连接蛋白 E-cadherin 的表达。CHOP 是内质网应激的重要标志物，其激活可促进一系列下游反应。研究表明，内质网应激会上调豚鼠胃黏膜细胞中 CHOP 的表达，而 CHOP 的激活可进一步上调人骨髓间充质干细胞中的黏着斑激酶（FAK）水平，从而激活小鼠神经母细胞瘤细胞中的 RhoA 活性。这一系列证据表明，内质网应激可能通过激活 RhoA 信号通路破坏肠道细胞间的结构完整性。然而，有关内质网应激是否参与芥酸破坏草鱼肠道细胞间结构完整性这一过程，目前还不清楚，有待研究。

近年来的研究表明，抗营养因子如棉酚能够诱导大菱鲆原代肌肉细胞发生内质网应激，这表明抗营养因子可能通过引发内质网应激，对鱼类健康产生负面影响。然而，芥酸与动物内质网应激之间的关系尚未有相关研究。已有证据表明，芥酸能引起大鼠心脏中游离脂肪酸（FFA）的增多，而FFA的增加又会导致小鼠胰腺β-TC3细胞系内质网应激的发生。这提示芥酸可能也能诱导动物体内的内质网应激，但这一假设仍需进一步验证。因此，本研究将通过系统的实验设计，深入探讨芥酸对草鱼肠道细胞间结构完整性的影响及其可能的机制，特别是内质网应激在其中的作用。这不仅有助于揭示芥酸对水产动物健康的潜在危害，也为改善水产养殖中菜籽粕等植物蛋白源的使用提供了理论依据。

4.1 材料与方法

根据研究目的，本部分需要开展2个试验：①芥酸对生长中期草鱼肠道内质网应激的影响；②内质网应激在芥酸破坏草鱼肠细胞间结构完整性中的作用。

4.1.1 芥酸对生长中期草鱼肠道内质网应激的影响

4.1.1.1 试验设计

同第二章。

4.1.1.2 试验饲料

同第二章。

4.1.1.3 试验条件和饲养管理

同第二章。

4.1.1.4 指标测定

（1）草鱼肠细胞超微结构观察。

草鱼肠细胞超微结构观察主要参照Song等（2013）的方法进行，并做适当修改。具体方法如下：首先将肠道组织经3%的戊二醛固定，然后用1%四氧化锇进行再固定。经丙酮逐级脱水后，先后经由脱水剂和Epon812环氧树脂比例（3:1、1:1、1:3）配成的不同浓度的渗透剂浸透，随后包埋，切片。将切好的超薄片先后经醋酸铀和枸橼酸铅染色，最后用JEM-1400PLUS透射电镜（日本电子株式会社，日本）观察草鱼肠细胞超微结构，并拍照。

(2) 基因表达。

操作方法同第二章，荧光定量引物设计主要根据 NCBI 上已公布的草鱼基因序列进行设计，采用 primer 6.0 进行引物设计，引物序列见表 4-1。

表 4-1 荧光定量引物

目标基因	正向引物（5′→3′）	反向引物（5′→3′）	温度/℃	登录号
GRP78	GTCACCTTTGAGATCGACGTG	AGAGAGTAGGCGTAGCTC	61.4	FJ436356
CHOP	GAATCCGAAACAGCCGAGGA	CCACACCTAGCACACCAGAC	64.5	KX013389
eIF2α	ATCAATAGCGGAGATGGGCG	TGATGACCACCACGCATTCA	61.4	KJ126860
XBP1	TTCTGAGTCCGCAGCAGGTG	GTTCTGGGTCAAGGATGTCC	55.0	KU509247
β-actin	GGCTGTGCTGTCCCTGTA	GGGCATAACCCTCGTAGAT	61.4	M25013

(3) 蛋白表达。

方法同第二章。抗体：GRP78（1∶1000，ABclonal，中国）、p-PERK（Thr^{982}）（1∶1000，Affinity BioReagents，美国）、p-IRE1（Ser^{724}）（1∶1000，Affinity BioReagents，美国）和 ATF6（1∶1000，ABclonal，中国）。

4.1.1.5 数据统计

同第二章。

4.1.2 内质网应激在芥酸破坏草鱼肠道细胞间结构完整性中的作用

4.1.2.1 试验设计

通过添加内质网应激抑制剂（4-PBA），考察内质网应激在芥酸破坏草鱼肠道细胞间结构完整性中的作用，试验共设计 4 个处理，每个处理 6 个重复，分别为对照组（Control）、内质网应激抑制剂组（4-PBA+）、芥酸组（EA+）和芥酸+内质网应激抑制剂组（EA+ 4-PBA+），见表 4-2。

表 4-2 试验设计

处理	对照组	对照组+抑制剂	处理组	处理组+抑制剂
重复数	6	6	6	6
DMSO	+	−	+	−

续表

处理	对照组	对照组+抑制剂	处理组	处理组+抑制剂
内质网应激抑制剂	-	+	-	+
EA	-	-	+	+

4.1.2.2 试验材料

（1）仪器与设备。

同第三章。

（2）试剂与药品。

4-PBA 购于 CSNpharm（芝加哥，美国）。其他试剂同第三章。

4.1.2.3 草鱼肠细胞培养

同第三章。

4.1.2.4 芥酸配制

同第三章。

4.1.2.5 内质网应激抑制剂配制

将内质网应激抑制剂 4-PBA 溶于 DMSO 中，得到浓度为 10 mmol/L 的储备液，-20 ℃条件下备用。使用前，按 1000 倍稀释。

4.1.2.6 细胞处理

将草鱼肠细胞接种于 12 孔板中培养 48h，随后添加 10 μmol/L 内质网应激抑制剂 4-PBA（预试验确定浓度）预处理 2h，再添加 3 mmol/L 芥酸（由第三章中筛选出的浓度）处理 24 h，阴性对照为不添加芥酸（0 mmol/L）。

4.1.2.7 观测指标

（1）草鱼肠细胞超微结构观察。

同上。

（2）基因表达。

内质网应激相关基因引物同上，紧密连接蛋白和黏附连接蛋白相关基因引物同第二章。基因表达所需试剂及耗材同第二章。

（3）内质网应激蛋白表达。

同上。

（4）RhoA 活性检测。

同第二章。

4.1.2.8 数据统计分析

本试验所有数据用平均值±标准误（Mean±SE）表示。对芥酸与内质网应激抑制剂的作用采用 SAS 软件（SAS Institute, Inc., 2006）的 2×2 多因素进行分析，并对处理组间的差异进行方差分析。$P \leqslant 0.05$ 为差异显著，$0.05 < P \leqslant 0.1$ 为差异显著趋势。

4.2 试验结果

4.2.1 芥酸对生长中期草鱼肠道细胞超微结构的影响

芥酸对生长中期草鱼肠道细胞超微结构的影响如表 4-3 和图 4-1 所示。0.00（对照组）、0.29%和 0.60%芥酸处理组中肠道细胞内质网排列整齐，呈正常的形态（扁平囊状），线粒体形状正常。当饲料中芥酸添加量达到 0.88%或以上时，草鱼肠道细胞中的内质网和线粒体会呈现出明显的肿胀现象。

表 4-3 芥酸对生长中期草鱼肠道细胞超微结构的影响

细胞器肿胀	饲料芥酸水平/（%饲料）					
	0.00	0.29	0.60	0.88	1.21	1.50
线粒体肿胀	-	-	-	++	++	+++
内质网肿胀	-	-	-	++	+++	+++

注 -表示无病理现象；+++表示病理现象严重；++表示病理现象轻微；+表示有极少的病理现象。

4.2.2 芥酸对生长中期草鱼肠道内质网应激相关基因表达的影响

芥酸对生长中期草鱼前、中和后肠内质网应激相关基因 mRNA 水平的影响如图 4-2 所示。随着饲料中芥酸水平的升高，草鱼肠道内质网应激相关指标 *GRP78*、*CHOP*、*eIF2α* 和 *XBP1* 的 mRNA 水平呈现出线性升高的变化（$P<0.01$）。

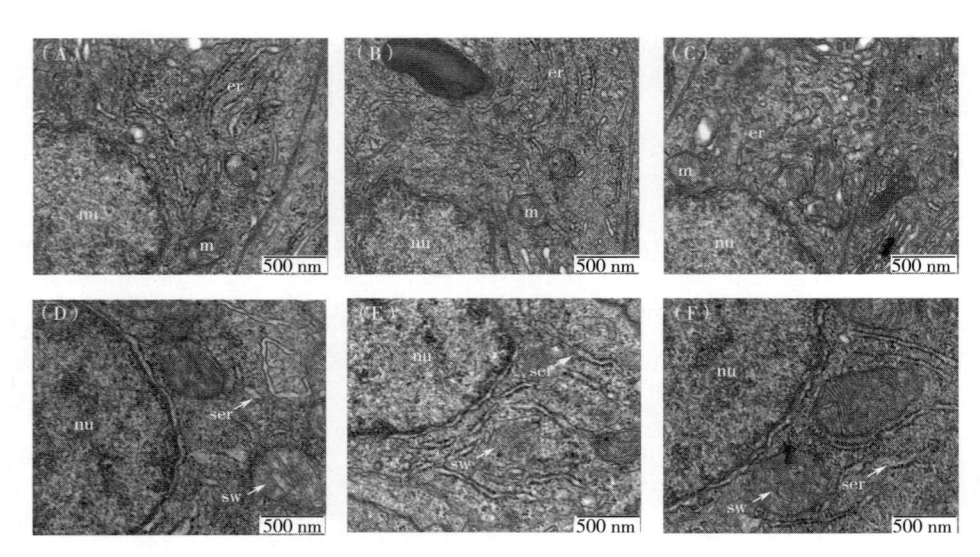

图4-1 芥酸对生长中期草鱼肠道细胞超微结构的影响（TEM，30000×）

（A）对照组　（B）0.29%芥酸组　（C）0.60%芥酸组　（D）0.88%芥酸组　（E）1.21%芥酸组　（F）1.50%芥酸组　缩写如下：细胞核：nu；内质网：er；内质网肿胀：ser；线粒体：m；线粒体肿胀：sw

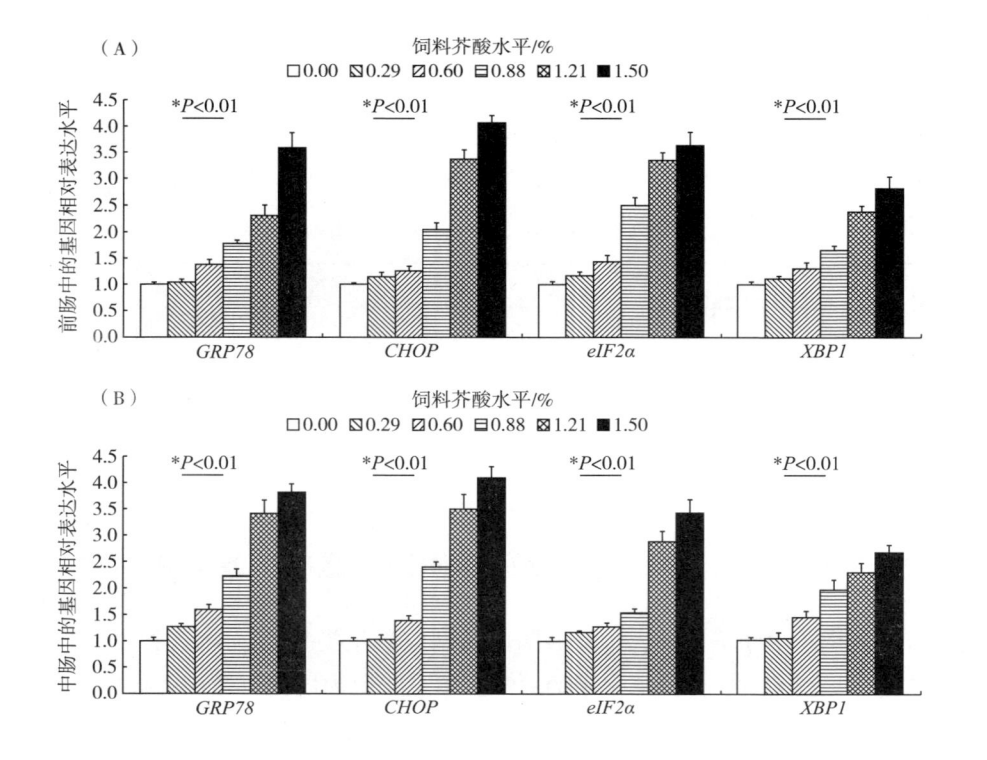

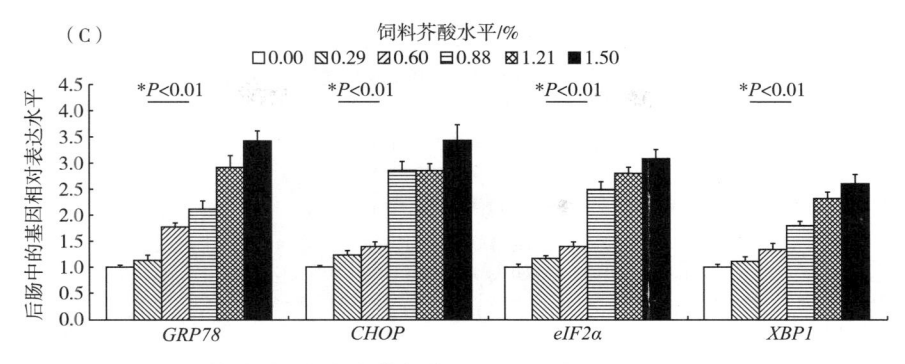

图 4-2 芥酸对生长中期草鱼前肠（A）、中肠（B）和后肠（C）
内质网应激相关基因 mRNA 水平的影响

数据表示为平均值±标准误（n=3，6 条鱼/组）；实线上方的 P 值表示呈现出明显的线性剂量关系（$P<0.01$）

4.2.3　芥酸对生长中期草鱼肠道内质网应激及非折叠蛋白反应关键蛋白表达的影响

芥酸对生长中期草鱼前、中和后肠内质网应激相关蛋白表达的影响如图 4-3 所示。随着饲料中芥酸水平的升高，草鱼肠道的内质网应激相关蛋白 GRP78、p-PERK、p-IRE1 和 ATF6 的蛋白表达呈现出线性升高的变化（$P<0.01$）。

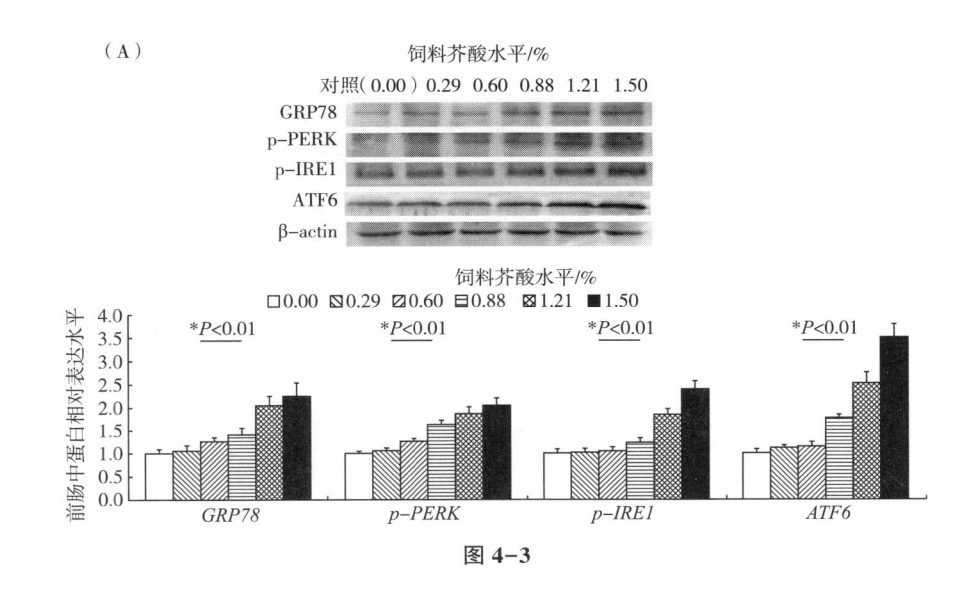

图 4-3

图 4-3 芥酸对生长中期草鱼前肠（A）、中肠（B）和
后肠（C）内质网应激相关蛋白表达的影响

数据表示为平均值±标准误（$n=3$，6 条鱼/组）；实线上方的 P 值
表示呈现出明显的线性剂量关系（$P<0.01$）

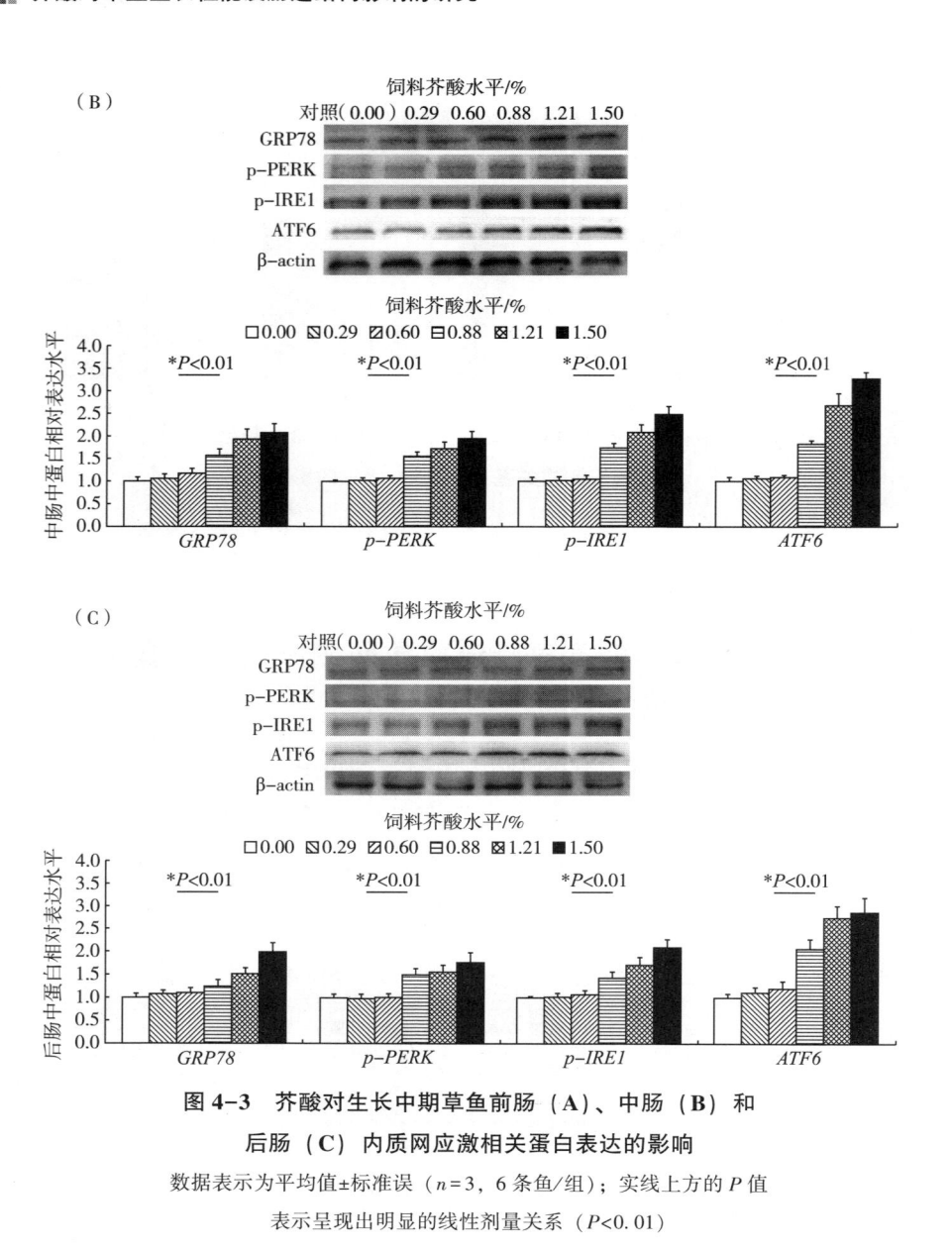

4.2.4 芥酸对草鱼肠细胞超微结构的影响

芥酸对草鱼肠细胞超微结构的影响如表 4-4 和图 4-4 所示。与对照组相比，芥酸会导致草鱼肠细胞内质网和线粒体呈现出明显的肿胀现象，而内质网应激抑制剂（4-PBA）可明显缓解由芥酸导致的草鱼肠细胞内质网和线粒体肿胀现象的发生。

表 4-4　芥酸和内质网应激抑制剂（4-PBA）处理
草鱼肠细胞后对草鱼肠细胞超微结构的影响

处理	对照组	对照组+抑制剂	处理组	处理组+抑制剂
线粒体肿胀	-	-	+++	++
内质网肿胀	-	-	++	+

注　-表示无病理现象；+++表示病理现象严重；++表示病理现象轻微；+表示有极少的病理现象。

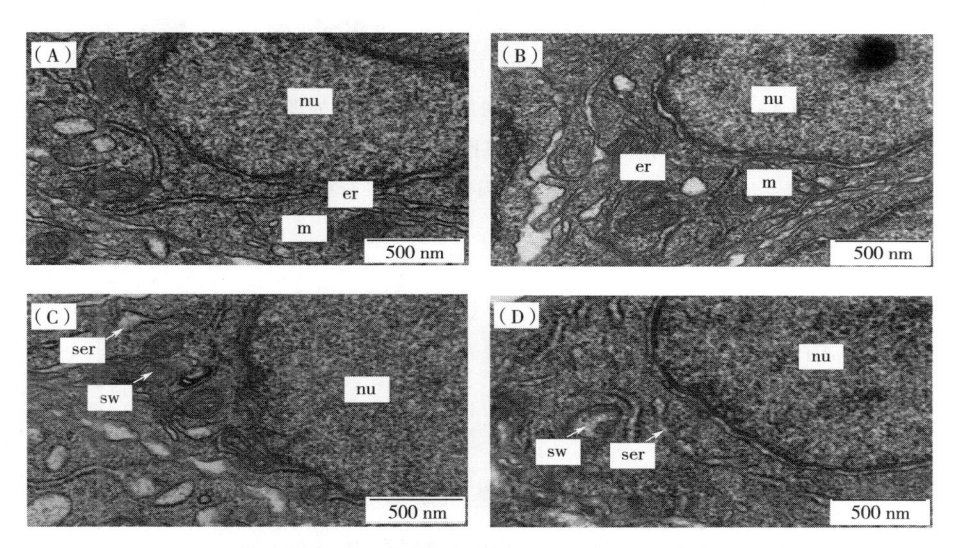

图 4-4　芥酸和内质网应激抑制剂（4-PBA）处理草鱼肠细胞后
对草鱼肠细胞超微结构的影响（TEM，30000×）

（A）对照组　（B）4-PBA 组　（C）芥酸组　（D）4-PBA+芥酸组　缩写如下：
细胞核，nu；内质网，er；内质网肿胀，ser；线粒体：m；线粒体肿胀：sw

4.2.5　芥酸对草鱼肠细胞内质网应激相关基因 mRNA 水平的影响

芥酸对草鱼肠细胞内质网应激相关基因 mRNA 水平如图 4-5 所示。与对照组相比，芥酸会显著上调草鱼肠细胞中 *GRP78*、*CHOP*、*eIF2α* 和 *XBP1* 的 mRNA 水平（$P<0.05$），而内质网应激抑制剂（4-PBA）可显著下调由芥酸上调的草鱼肠道细胞 *GRP78*、*CHOP*、*eIF2α* 和 *XBP1* 的 mRNA 水平（$P<0.01$）。

4.2.6　芥酸对草鱼肠细胞内质网应激相关蛋白表达的影响

芥酸对草鱼肠细胞内质网应激相关蛋白表达的影响如图 4-6 所示。与对照组

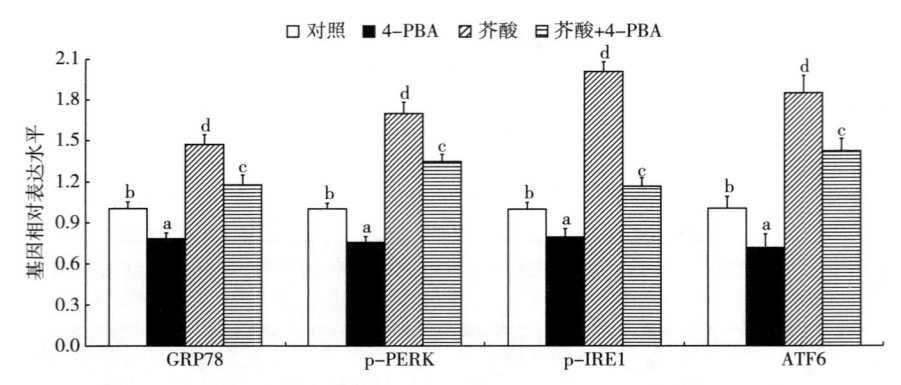

图 4-5 芥酸和内质网应激抑制剂（4-PBA）处理草鱼肠细胞后
对内质网应激相关基因 mRNA 水平的影响

数据表示为平均值±标准误（$n=6$）；图柱上的不同字母表示显著差异（$P<0.05$）

相比，芥酸会显著上调草鱼肠细胞中 GRP78、p-PERK、p-IRE1 和 ATF6 的蛋白表达（$P<0.05$），而内质网应激抑制剂（4-PBA）可显著下调由芥酸上调的草鱼肠道细胞 GRP78、p-PERK、p-IRE1 和 ATF6 的蛋白表达（$P<0.05$）。

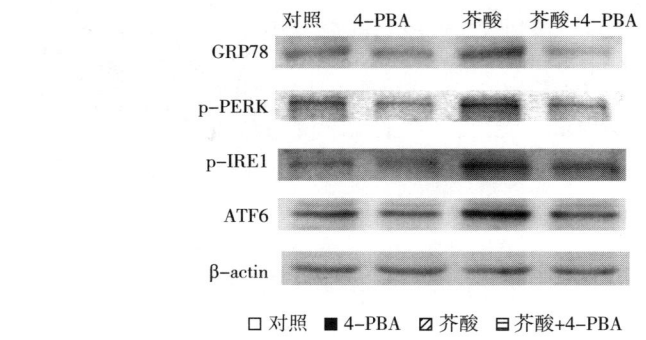

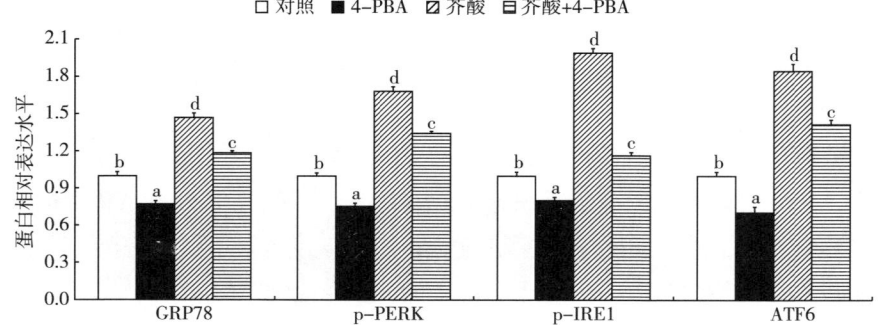

图 4-6 芥酸和内质网应激抑制剂（4-PBA）处理草鱼肠
细胞后对内质网应激相关蛋白表达的影响

数据表示为平均值±标准误（$n=6$）；图柱上的不同字母表示显著差异（$P<0.05$）

芥酸对草鱼肠细胞 RhoA 活性的影响如图 4-7 所示。与对照组相比，芥酸可显著增加草鱼肠细胞中 GTP-RhoA/Total-RhoA 比值（$P<0.05$），而内质网应激抑制剂（4-PBA）可显著降低由芥酸增加的草鱼肠道细胞中 GTP-RhoA/Total-RhoA 比值（$P<0.05$）。

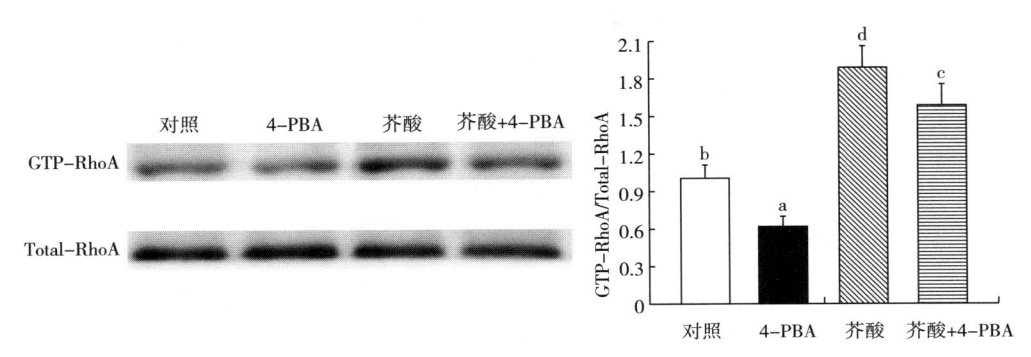

图 4-7　芥酸和内质网应激抑制剂（4-PBA）处理草鱼肠细胞后对草鱼肠细胞 GTP-RhoA/Total-RhoA 比值的影响

数据表示为平均值±标准误（$n=6$）；图柱上的不同字母表示显著差异（$P<0.05$）

芥酸对草鱼肠细胞间紧密连接蛋白和黏附连接蛋白相关基因 mRNA 水平的影响如图 4-8 所示。与对照组相比，芥酸下调草鱼肠细胞间紧密连接蛋白 $ZO-1$、$occludin$、$claudin-7a$ 和 $claudin-7b$（$P<0.05$），以及黏附连接蛋白 $nectin$、$E-cadherin$ 和 $\beta-catenin$ 的 mRNA 水平（$P<0.05$），同时上调离子通道蛋白 $claudin-15a$ 以及关键信号分子 $MLCK$、$NMII$ 和 $ROCK$ 的 mRNA 水平（$P<0.05$），而内质网应激抑制剂（4-PBA）可显著上调由芥酸下调的 $ZO-1$、$occludin$、$nectin$ 和 $E-cadherin$ 的 mRNA 水平（$P<0.05$），同时下调由芥酸上调的 $claudin-15a$、$MLCK$、$NMII$ 和 $ROCK$ 的 mRNA 水平（$P<0.05$）。与此同时，而内质网应激抑制剂（4-PBA）有上调由芥酸下调的 $claudin-7a$（$P=0.053$）、$claudin-7b$（$P=0.055$）和 $\beta-catenin$（$P=0.087$）的 mRNA 水平的趋势。

芥酸对草鱼肠细胞间 E-cadherin、β-catenin 和 ZO-1 蛋白表达的影响如图 4-9 所示。与对照组相比，芥酸可显著下调草鱼肠细胞间 E-cadherin、β-catenin 和 ZO-1 的蛋白表达（$P<0.05$），而内质网应激抑制剂（4-PBA）可显著上调由芥酸下调的草鱼肠道细胞 E-cadherin、β-catenin 和 ZO-1 的蛋白表达（$P<0.05$）。

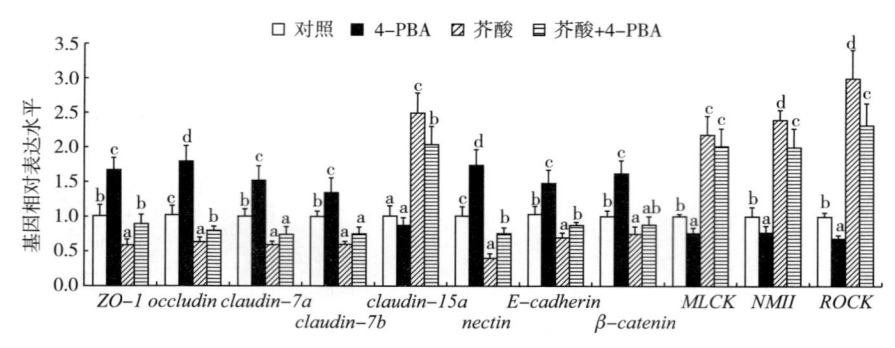

图 4-8 芥酸和内质网应激抑制剂（4-PBA）处理草鱼肠细胞后对草鱼肠细胞间紧密连接和黏附连接相关基因 mRNA 水平的影响

数据表示为平均值±标准误（$n=6$）；图柱上的不同字母表示显著差异（$P<0.05$）

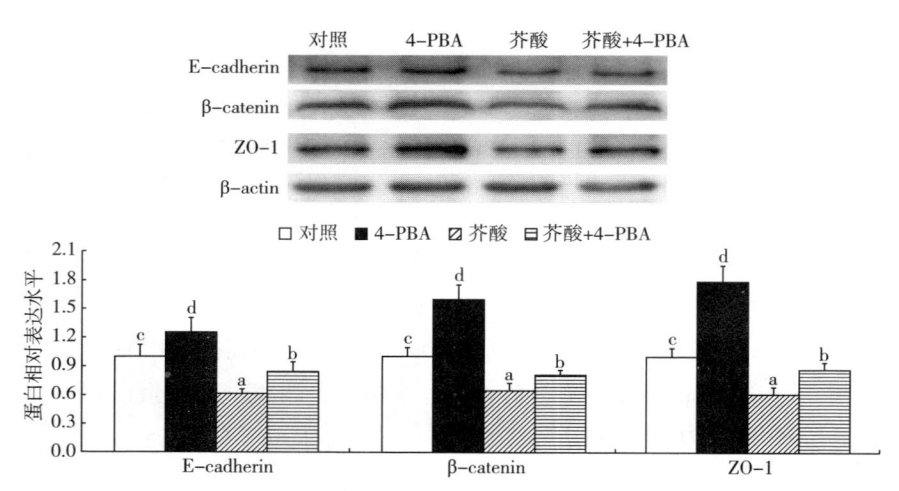

图 4-9 芥酸和内质网应激抑制剂（4-PBA）处理草鱼肠细胞后对 E-cadherin、β-catenin 和 ZO-1 蛋白表达的影响

数据表示为平均值±标准误（$n=6$）；图柱上的不同字母表示显著差异（$P<0.05$）

4.3 讨论

4.3.1 芥酸对生长中期草鱼肠道内质网应激的影响

内质网应激是动物应对外界刺激的一种适应性机制，然而当环境刺激持续时间过长，动物体内质网应激的发生便会对动物产生负面影响。据报道，细胞中内

质网肿胀是内质网应激发生的标志。本试验的结果表明：芥酸会导致生长中期草鱼肠道细胞内质网肿胀现象的发生。不利因素引起动物细胞内质网肿胀的现象在其他研究中也有报道：农药吡虫啉和氧化锌纳米粒子都会导致小鼠肝脏细胞内质网肿胀的发生以及薯蓣皂苷会导致人肝癌细胞内质网肿胀的发生。由此可见：芥酸会导致生长中期草鱼肠道内质网应激的发生。GRP78 和 CHOP 是内质网应激发生的关键指标。在本研究中，芥酸会显著上调生长中期草鱼前、中和后肠三个肠段中 *GRP78* 和 *CHOP* 的基因表达，以及上调 GRP78 的蛋白表达，这也说明芥酸会导致生长中期草鱼肠道内质网应激的发生。非折叠蛋白反应是内质网应激介导的下游信号途径，其主要包含三条信号通路，分别是：PERK/eIF2α、IRE1/XBP1 和 ATF6。当内质网腔中非折叠蛋白或错误折叠蛋白蓄积时，内质网应激的发生会导致这三个跨膜受体与 GRP78 分开，进而激活 UPR 信号途径。本试验的研究结果表明：芥酸会增加生长中期草鱼前、中和后肠三个肠段中 p-PERK、p-IRE1 和 ATF6 的蛋白表达，这进一步说明芥酸会诱导生长中期草鱼肠道内质网应激。研究表明：棉酚会导致人白血病细胞株、人肝癌细胞、大菱鲆原代肌肉细胞发生内质网应激以及单宁会导致人前列腺上皮细胞发生内质网应激。这说明抗营养因子确实会诱导动物内质网应激的发生，这可能是抗营养因子对动物产生负面影响的方式之一。其他不利因素也会诱导动物内质网应激。例如，小鼠上的研究发现，呕吐毒素会导致其肠道发生内质网应激以及高脂会导致其结肠发生内质网应激。在人上的研究表明，T-2 毒素会导致其结肠癌细胞 Caco-2 内质网应激的发生以及慢病毒感染会导致其肠道内质网应激的发生。这些研究结果说明，内质网应激的发生可能是外界不利因素对动物造成负面影响的方式之一。在本试验中，进一步的相关性（表 4-5）分析结果显示 p-PERK、p-IRE1 和 ATF6 蛋白表达水平与 GRP78 蛋白表达水平呈正相关，说明芥酸可能会通过上调 GRP78 的蛋白表达（内质网应激）进而上调 p-PERK、p-IRE1 和 ATF6 蛋白表达（非折叠蛋白反应）。芥酸诱导草鱼肠道内质网应激的发生可能跟其导致鱼体内游离脂肪酸含量的增多有关。Nivala 等（2013）研究发现游离脂肪酸增多会诱导大鼠胰腺 β-细胞发生内质网应激。Pasini 等（1992）研究发现芥酸会导致大鼠体内游离脂肪酸的增多。综上，我们猜测芥酸可能会导致鱼体内游离脂肪酸增多，进而诱导内质网应激。然而，该假设还有待进一步验证。另外，芥酸诱导草鱼肠道内质网应激的发生可能跟其抑制其他类型的脂肪酸代谢有关。Christophersen 等（1972）研究发现芥酸会抑制大鼠心脏中脂肪酸的 β-氧化过程，这可能会导致各种类型的脂肪酸在动物体内蓄积，比如棕榈酸。

而近年来，大量的研究表明：棕榈酸会导致大鼠肝癌细胞、大鼠胰腺β-细胞和人肝细胞内质网应激。因此，我们猜测芥酸可能会导致鱼体内游离脂肪酸含量的增多或棕榈酸的蓄积，进而诱发鱼体内质网应激的发生。然而该假设也有待进一步验证。

内质网应激的发生会破坏动物肠道物理屏障功能。第三章中的研究结果显示：芥酸可通过激活RhoA信号途径破坏草鱼肠道细胞间结构完整性。同时，在本试验中，我们发现芥酸会诱导生长中期草鱼肠道内质网应激。因此，我们猜测芥酸破坏草鱼肠道细胞间结构完整性可能跟其诱导内质网应激的发生有关。接下来我们开展细胞试验探讨内质网应激在芥酸破坏草鱼肠道细胞间结构完整性中的作用。

表4-5 生长中期草鱼肠道内质网应激指标相关性分析

自变量	因变量	前肠		中肠		后肠	
		相关系数	P	相关系数	P	相关系数	P
GRP78 mRNA 水平	GRP78 蛋白水平	+0.953	<0.01	+0.994	<0.01	+0.941	<0.01
GRP78 蛋白水平	p-PERK 蛋白水平	+0.963	<0.01	+0.991	<0.01	+0.900	<0.05
	p-IRE1 蛋白水平	+0.970	<0.01	+0.989	<0.01	+0.978	<0.01
	ATF6 蛋白水平	+0.973	<0.01	+0.986	<0.01	+0.905	<0.05
p-PERK 蛋白水平	$eIF2\alpha$	+0.995	<0.01	+0.935	<0.01	+0.987	<0.01
p-IRE1 蛋白水平	$XBP1$	+0.977	<0.01	+0.980	<0.01	+0.987	<0.01

4.3.2 内质网应激在芥酸破坏草鱼肠道细胞间结构完整性中的作用

为探究内质网应激在芥酸破坏草鱼肠道细胞间结构完整性中的作用，我们进一步通过细胞试验进行了验证。在本研究中，芥酸会下调 ZO-1、occludin、claudin-7a、claudin-7b、nectin、E-cadherin 和 β-catenin 的基因表达，上调 claudin-15a 的基因表达，和下调 ZO-1、E-cadherin 和 β-catenin 的蛋白表达，以及上调内质网应激相关指标 GRP78、CHOP、eIF2α 和 XBP1 的基因表达，和内质网应激关键蛋白 GRP78、p-PERK、p-IRE1 和 ATF6 的蛋白表达，这说明芥酸可能会通过诱导草鱼肠细胞内质网应激破坏其细胞间结构完整性。除此之外，我们实验室前期的研究结果发现棉酚会上调 MLCK 信号分子，进而破坏草鱼肠细胞间结构完整性。本试验的研究结果显示：芥酸会上调草鱼肠细胞 MLCK 的基因表达。在人类方面的研究表明：内质网应激会上调其血管平滑肌细胞 MLCK 的蛋白表达，这也说明芥酸会诱导草鱼肠细胞内质网应激的发生破坏其细胞间结构完整性。4-PBA 是内质网应激的特异性抑制剂，其被广泛应用于内质网应激相关的生物学功能研究中。在本研究中，与单独添加芥酸相比，在芥酸破坏草鱼肠细胞间结构完整性前添加内质网应激抑制剂 4-PBA 可显著上调 *ZO-1*、*occludin*、*nectin* 和 *E-cadherin* 的基因表达，下调 *claudin-15a* 的基因表达，和上调 ZO-1、E-cadherin 和 β-catenin 的蛋白表达。这说明芥酸会诱导草鱼肠细胞内质网应激，进而破坏草鱼肠细胞间结构完整性。鼠上的研究表明：内质网应激会下调肠道中 ZO-1 和 occludin 的蛋白表达和脊髓中 β-catenin 和 occludin 的蛋白表达。人类方面的研究表明：内质网应激会下调结肠癌细胞 Caco-2 中 E-cadherin、β-catenin 和 ZO-1 的蛋白表达。由此可见，内质网应激的发生会导致动物肠道细胞间结构完整性的破坏，同时，芥酸可以通过诱导草鱼肠道内质网应激进而破坏其肠道细胞间结构完整性。除此之外，在本研究中，芥酸会激活 RhoA 信号分子，而内质网应激抑制剂会明显抑制由芥酸激活的 RhoA 信号分子。这说明芥酸可诱导草鱼肠细胞内质网应激进而激活 RhoA 信号分子。内质网应激激活 RhoA 信号分子的现象在其他试验中也有发现。例如，Liang 等（2013）研究发现内质网应激会增加小鼠血管平滑肌细胞中 RhoA 的活性。同时，在大鼠 FR3T3 细胞中，内质网应激会激活 RhoA 信号分子。综上所述：本试验进一步通过体外试验证明芥酸会诱导内质网应激的发生，进而激活 RhoA 信号途径，最终破坏草鱼肠道细胞间结构完整性。

4.4 小结

根据试验结果，得出以下结论：

（1）芥酸会引起草鱼肠道和肠细胞内质网应激。

（2）芥酸破坏草鱼肠细胞间结构完整性，这与芥酸引起内质网应激，激活 RhoA 信号途径，上调关键信号分子 *MLCK*、*ROCK* 和 *NMII* 的基因表达，下调紧密连接（*ZO-1*、*occludin*、*claudin-7a* 和 *-7b*）和黏附连接蛋白（*E-cadherin*、*β-catenin* 和 *nectin*）的基因表达，上调离子通道蛋白 *claudin-15a* 的基因表达，以及降低 E-cadherin、β-catenin 和 ZO-1 的蛋白表达有关。

第五章　内质网应激介导的非折叠蛋白反应

第四章的研究结果表明，内质网应激在芥酸破坏草鱼肠道细胞间结构完整性过程中发挥了重要作用。然而，内质网应激究竟是通过哪条信号途径参与这一过程，目前尚不明确。已有研究表明，内质网应激对动物产生的负面影响主要是通过激活其下游的非折叠蛋白反应来实现的。非折叠蛋白反应主要包括三条信号途径：PERK/eIF2α、IRE1/XBP1 和 ATF6。相关研究显示，PERK 信号途径的激活会降低大鼠蛛网膜中紧密连接蛋白 ZO-1 和 occludin 的蛋白表达；郑晨果（2017）在人肠道的研究中发现，激活 IRE1 信号途径会降低黏附连接蛋白 E-cadherin 的蛋白表达。此外，ATF6 信号途径的激活也会导致小鼠枯否细胞 KCs 中黏附连接蛋白 β-catenin 的表达降低，并下调人肾皮质近曲小管上皮细胞中黏附连接蛋白 E-cadherin 和紧密连接蛋白 ZO-1 的蛋白表达。这些研究表明，非折叠蛋白反应的激活可能在芥酸引起的草鱼肠道细胞间结构完整性损伤中起到了重要作用。细胞间的紧密连接蛋白和黏附连接蛋白受 RhoA 信号途径的调控。在人类方面的研究中，激活 IRE1 信号途径可增强结肠癌细胞 HCT116 中 RhoA 的活性；而在另一个实验中，激活 ATF4（PERK 的下游信号分子）也能够增加大鼠海马原代神经元中 RhoA 的活性。这些证据表明，内质网应激介导的非折叠蛋白反应可能是芥酸诱导草鱼肠道内质网应激并破坏其细胞间结构完整性的关键途径。然而，目前关于芥酸与非折叠蛋白反应相关信号途径的研究尚未有明确报道，因此有必要进一步探索芥酸与非折叠蛋白反应信号途径之间的关系。

研究发现，非折叠蛋白反应中的三条主要信号途径（PERK/eIF2α、IRE1/XBP1 和 ATF6）并非在所有外界刺激引发的负面效应中都会发挥作用。例如，在斑马鱼肝细胞中，棕榈酸通过激活 PERK/eIF2α 和 IRE1/XBP1 信号途径而非 ATF6 信号途径对其产生负面效应。Fan 等（2014）在大鼠肠上皮细胞中的研究也表明，厄洛替尼通过激活 PERK/eIF2α 和 IRE1/XBP1 信号途径而非 ATF6 信号途径下调 E-cadherin 的表达。此外，针对金属元素的研究也发现，镉和锌能分别上调三倍体湘云鲫和黄颡鱼肝脏中的 PERK 和 IRE1 基因表达，但对 ATF6 表达没有影响。然而，人类研究表明，晚期氧化蛋白产物降低近端肾小管细胞中 ZO-1 和 E-cadherin 的表达是通过激活 ATF6 信号途径，而非 PERK 和 IRE1 途径实现的。这些差异表明，不同刺激对动物的负面影响可能通过不同的信号途径发挥作用。对于芥酸是否通过激活 PERK、IRE1 和 ATF6 信号途径破坏草鱼肠道细胞间结构完整性，仍然缺乏足够的证据。

因此，本研究计划通过细胞实验，分别阻断非折叠蛋白反应中的三条信号途径，系统研究这三条信号途径在芥酸破坏草鱼肠道细胞间结构完整性过程中的具体作用。通过这一系列实验，旨在揭示芥酸对水产动物肠道屏障功能的潜在危害，为水产养殖中植物蛋白源的安全性评价提供科学依据，并为未来的研究提供新的方向。

5.1 材料与方法

根据研究目的，本部分共需要开展 3 个细胞试验：①PERK/eIF2α 信号途径在芥酸影响草鱼肠细胞间结构完整性中的作用；②IRE1/XBP1 信号途径在芥酸影响草鱼肠细胞间结构完整性中的作用；③ATF6 信号途径在芥酸影响草鱼肠细胞间结构完整性中的作用。

5.1.1 PERK/eIF2α 信号途径在芥酸影响草鱼肠细胞间结构完整性中的作用

5.1.1.1 试验设计

通过添加 PERK 抑制剂（GSK2656157），考察 PERK/eIF2α 信号途径在芥酸破坏草鱼肠细胞间结构完整性中的作用，试验共设计 4 个处理，每个处理 6 个重复，分别为对照组（control）、PERK 抑制剂组（GSK2656157+）、芥酸组（EA+）和芥酸+PERK 抑制剂组（EA+GSK2656157+），见表 5-1。

表 5-1 试验设计 1

处理	对照组	对照组+抑制剂	处理组	处理组+抑制剂
重复数	6	6	6	6
DMSO	+	−	+	−
PERK 抑制剂	−	+	−	+
EA	−	−	+	+

5.1.1.2　试验材料

（1）仪器与设备。
同第三章。
（2）试剂与药品。
PERK 活性抑制剂（GSK2656157）购于 CSNpharm（芝加哥，美国）。其他试剂同第三章。

5.1.1.3　草鱼肠细胞培养

同第三章。

5.1.1.4　芥酸配制

同第三章。

5.1.1.5　PERK 活性抑制剂配制

将 PERK 活性抑制剂 GSK2656157 溶于 DMSO 中，得到浓度为 10 mmol/L 的储备液，-20℃条件下备用。使用前，按 1000 倍稀释。

5.1.1.6　细胞处理

将草鱼肠细胞接种于 12 孔板中培养 48 h，随后添加 10 μmol/L 抑制剂 GSK2656157 预处理 2 h，再用 3 mmol/L 芥酸（由第三章中确定的芥酸浓度）处理草鱼肠细胞 24 h。

5.1.1.7　观测指标

（1）RhoA 活性检测。
同第二章。
（2）基因表达。
内质网应激相关基因引物同第四章，紧密连接蛋白和黏附连接蛋白相关基因引物同第二章。基因表达所需试剂及耗材同第二章。
（3）蛋白表达。
同第二章和第四章。

5.1.1.8 数据统计分析

本试验所有数据用平均值±标准误（Mean±SE）表示。对芥酸与PERK抑制剂的作用采用SAS软件（SAS Institute，Inc.，2006）的2×2多因素进行分析，并对处理组间的差异进行方差分析。$P \leqslant 0.05$ 为差异显著，$0.05 < P \leqslant 0.1$ 为差异显著趋势。

5.1.2 IRE1/XBP1信号途径在芥酸破坏草鱼肠细胞间结构完整性中的作用

5.1.2.1 试验设计

通过添加IRE1抑制剂（STF-083010），考察IRE1/XBP1信号途径在芥酸破坏草鱼肠细胞间结构完整性中的作用，试验共设计4个处理，每个处理6个重复，分别为对照组（control）、IRE1抑制剂组（STF-083010+）、芥酸组（EA+）和芥酸+IRE1抑制剂组（EA+ STF-083010+），见表5-2。

表5-2　试验设计2

处理	对照组	对照组+抑制剂	处理组	处理组+抑制剂
重复数	6	6	6	6
DMSO	+	−	+	−
IRE1 抑制剂	−	+	−	+
EA	−	−	+	+

5.1.2.2 试验材料

（1）仪器与设备。

同第三章。

（2）试剂与药品。

STF-083010购于CSNpharm（芝加哥，美国）。其他试剂同第三章。

5.1.2.3 草鱼肠细胞培养

同第三章。

5.1.2.4 芥酸配制

同第三章。

5.1.2.5 IRE1 抑制剂配制

将 STF-083010 溶于 DMSO 中，得到浓度为 50 mmol/L 的储备液，-20℃ 条件下备用。使用前，按 1000 倍稀释。

5.1.2.6 细胞处理

将草鱼肠细胞接种于 12 孔板中培养 48 h，随后添加 50 μmol/L 抑制剂 STF-083010（预试验确定浓度）预处理 2 h，最后用 3 mmol/L 芥酸（由第三章中确定的芥酸浓度）处理草鱼肠细胞 24 h。

5.1.2.7 观测指标

（1）基因表达。

内质网应激相关基因引物同第四章，紧密连接蛋白和黏附连接蛋白相关基因引物同第二章，基因表达所需试剂及耗材同第二章。

（2）RhoA 活性检测。

同第二章。

（3）蛋白表达。

同第二章和第四章。

5.1.2.8 数据统计分析

本试验所有数据用平均值±标准误（Mean±SE）表示。对芥酸与 IRE1 抑制剂的作用采用 SAS 软件（SAS Institute, Inc., 2006）的 2×2 多因素进行分析，并对处理组间的差异进行方差分析。$P \leq 0.05$ 为差异显著，$0.05 < P \leq 0.1$ 为差异显著趋势。

5.1.3 ATF6 信号途径在芥酸破坏草鱼肠细胞间结构完整性中的作用

5.1.3.1 试验设计

通过添加 ATF6 抑制剂（AEBSF），考察 ATF6 信号途径在芥酸破坏草鱼肠细胞间结构完整性中的作用，试验共设计 4 个处理，每个处理 6 个重复，分别为对照组（control）、ATF6 抑制剂组（AEBSF+）、芥酸组（EA+）和芥酸+ATF6 抑制剂组（EA+ AEBSF+），见表 5-3。

表 5-3 试验设计 3

处理	对照组	对照组+抑制剂	处理组	处理组+抑制剂
重复数	6	6	6	6
DMSO	+	−	+	−
ATF6 抑制剂	−	+	−	+
EA	−	−	+	+

5.1.3.2　试验材料

（1）仪器与设备。

同第三章。

（2）试剂与药品。

AEBSF 购于 CSNpharm（芝加哥，美国）。其他试剂同第三章。

5.1.3.3　草鱼肠细胞培养

同第三章。

5.1.3.4　芥酸配制

同第三章。

5.1.3.5　ATF6 抑制剂配制

将 ATF6 抑制剂（AEBSF）溶于 DMSO 中，得到浓度为 300 mmol/L 的储备液，−20℃条件下备用。使用前，按 1000 倍稀释。

5.1.3.6　细胞处理

将草鱼肠细胞接种于 12 孔板中培养 48 h，随后添加 300 μmol/L 抑制剂 AEBSF（预试验确定浓度）处理 2 h，最后用 3 mmol/L 芥酸（由第三章中确定的芥酸浓度）处理草鱼肠细胞 24 h。

5.1.3.7 观测指标

（1）基因表达。

内质网应激相关基因引物同第四章，紧密连接蛋白和黏附连接蛋白相关基因引物同第二章，基因表达所需试剂及耗材同第二章。

（2）RhoA 活性检测。

同第二章。

（3）蛋白表达。

同第二章和第四章。

5.1.3.8 数据统计分析

本试验所有数据用平均值±标准误（Mean±SE）表示。对芥酸与 ATF6 抑制剂的作用采用 SAS 软件（SAS Institute，Inc.，2006）的 2×2 多因素进行分析，并对处理组间的差异进行方差分析。$P \leqslant 0.05$ 为差异显著，$0.05<P \leqslant 0.1$ 为差异显著趋势。

5.2 试验结果

5.2.1 PERK/eIF2α 信号途径在芥酸破坏草鱼肠细胞间结构完整性中的作用

芥酸对草鱼肠细胞 $eIF2\alpha$ 的 mRNA 水平的影响见图 5-1。与对照组相比，芥酸会显著上调草鱼肠细胞中 $eIF2\alpha$ 的 mRNA 水平（$P<0.05$），而 PERK 抑制剂（GSK2656157）可显著下调由芥酸上调的草鱼肠道细胞中 $eIF2\alpha$ 的 mRNA 水平（$P<0.05$）。

芥酸对草鱼肠细胞中 p-PERK 蛋白表达的影响见图 5-2。与对照组相比，芥酸会显著上调草鱼肠细胞中 p-PERK 的蛋白表达（$P<0.05$），而 PERK 抑制剂（GSK2656157）可显著下调由芥酸上调的草鱼肠道细胞中 p-PERK 的蛋白表达（$P<0.05$）。

芥酸和 PERK 抑制剂（GSK2656157）对草鱼肠细胞中 RhoA 活性的影响如图 5-3 所示。与对照组相比，芥酸会显著增加草鱼肠细胞中 GTP-RhoA/Total-

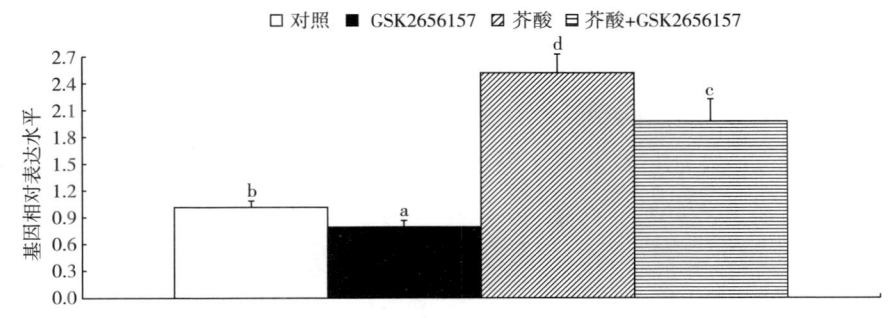

图 5-1 芥酸和 PERK 抑制剂（GSK2656157）处理草鱼肠细胞后对 *eIF2α* 的 mRNA 水平的影响

数据表示为平均值±标准误（$n=6$）；图柱上的不同字母表示显著差异（$P<0.05$）

RhoA 比值（$P<0.05$），而 PERK 抑制剂（GSK2656157）可显著降低由芥酸增加的草鱼肠道细胞中 GTP-RhoA/Total-RhoA 比值（$P<0.05$）。

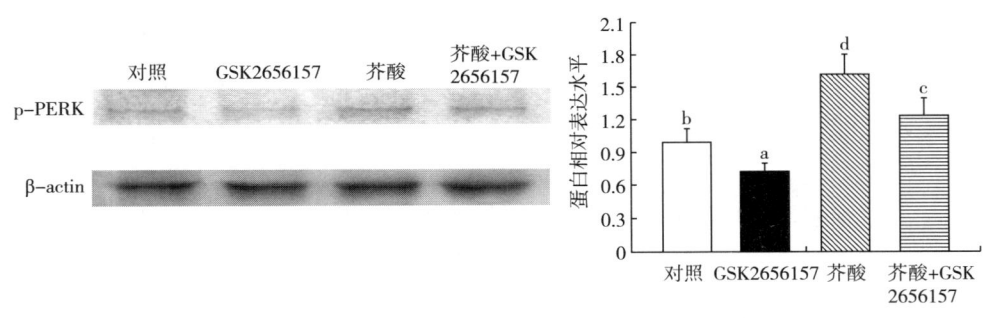

图 5-2 芥酸和 PERK 抑制剂（GSK2656157）处理草鱼肠细胞后对 p-PERK 蛋白表达的影响

数据表示为平均值±标准误（$n=6$）；图柱上的不同字母表示显著差异（$P<0.05$）

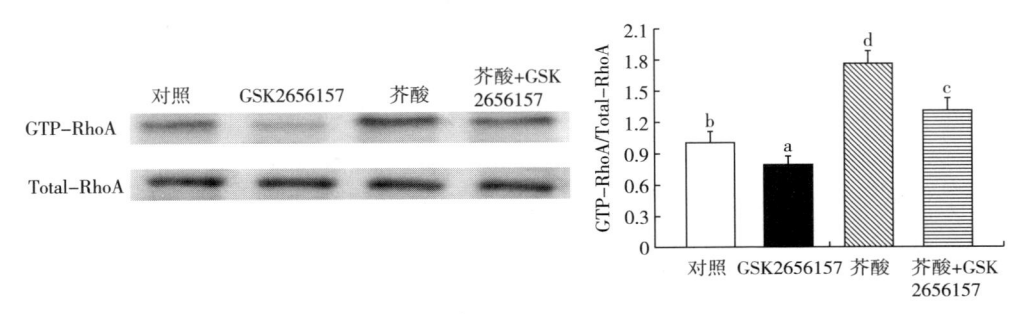

图 5-3 芥酸和 PERK 抑制剂（GSK2656157）处理草鱼肠细胞后对 GTP-RhoA/Total-RhoA 比值的影响

数据表示为平均值±标准误（$n=6$）；图柱上的不同字母表示显著差异（$P<0.05$）

芥酸对草鱼肠细胞间紧密连接蛋白和黏附连接蛋白相关基因 mRNA 水平的影响如图 5-4 所示。与对照组相比，芥酸下调草鱼肠细胞中紧密连接蛋白 $ZO-1$、$occludin$、$claudin-7a$ 和 $claudin-7b$（$P<0.05$），以及黏附连接蛋白 $nectin$、$E-cadherin$ 和 $\beta-catenin$ 的 mRNA 水平（$P<0.05$），同时上调离子通道蛋白 $claudin-15a$ 以及关键信号分子 $MLCK$、$NMII$ 和 $ROCK$ 的 mRNA 水平（$P<0.05$），而 PERK 抑制剂（GSK2656157）可显著上调由芥酸下调的 $ZO-1$、$claudin-7a$ 和 $claudin-7b$ 的 mRNA 水平（$P<0.05$），同时下调由芥酸上调的 $claudin-15a$、$MLCK$、$NMII$ 和 $ROCK$ 的 mRNA 水平（$P<0.05$）。与此同时，PERK 抑制剂（GSK2656157）有上调由芥酸下调的 $nectin$（$P=0.067$）mRNA 水平的趋势。

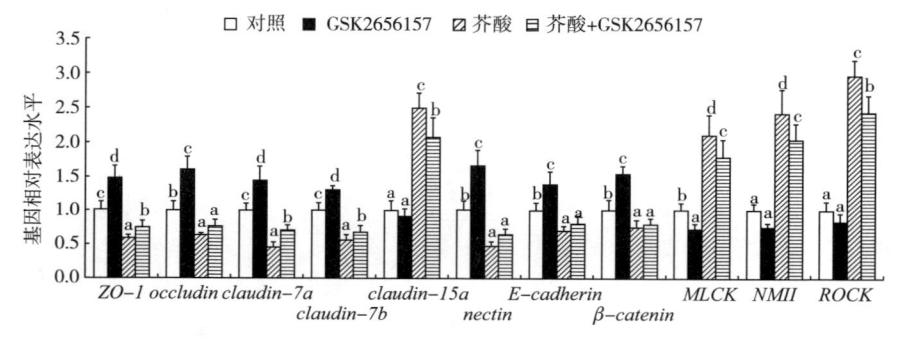

图 5-4 芥酸和 PERK 抑制剂（GSK2656157）处理草鱼肠细胞后对草鱼肠细胞间紧密连接蛋白和黏附连接蛋白相关基因 mRNA 水平的影响

数据表示为平均值±标准误（$n=6$）；图柱上的不同字母表示显著差异（$P<0.05$）

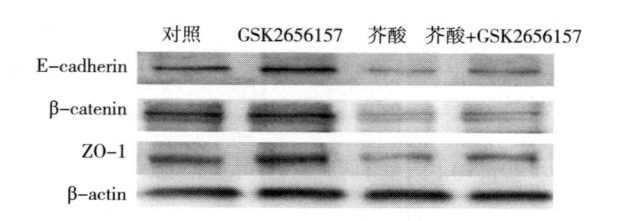

芥酸和 PERK 抑制剂（GSK2656157）对草鱼肠细胞中 E-cadherin、β-catenin 和 ZO-1 蛋白表达的影响如图 5-5 所示。与对照组相比，芥酸会显著下调草鱼肠细胞 E-cadherin、β-catenin 和 ZO-1 的蛋白表达（$P<0.05$），而 PERK 抑制剂（GSK2656157）可显著上调由芥酸下调的草鱼肠道细胞中 E-cadherin、β-catenin 和 ZO-1 的蛋白表达（$P<0.05$）。

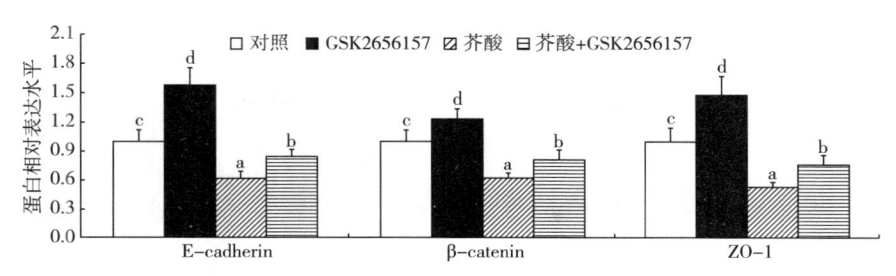

图 5-5 芥酸和 PERK 抑制剂（GSK2656157）处理草鱼肠细胞后对
E-cadherin、β-catenin 和 ZO-1 蛋白表达的影响

数据表示为平均值±标准误（$n=6$）；图柱上的不同字母表示显著差异（$P<0.05$）

5.2.2 IRE1/XBP1 信号途径在芥酸破坏草鱼肠细胞间结构完整性中的作用

芥酸和 IRE1 抑制剂（STF-083010）对草鱼肠细胞中 *XBP1* 的 mRNA 水平的影响如图 5-6 所示。与对照组相比，芥酸会显著上调草鱼肠细胞中 *XBP1* 的 mRNA 水平（$P<0.05$），而 IRE1 抑制剂（STF-083010）可显著下调由芥酸上调的草鱼肠道细胞中 *XBP1* 的 mRNA 水平（$P<0.05$）。

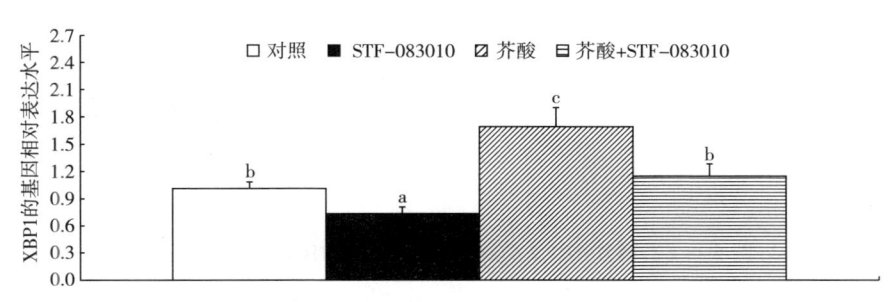

图 5-6 芥酸和 IRE1 抑制剂（STF-083010）处理草鱼肠
细胞后对 *XBP1* 的 mRNA 水平的影响

数据表示为平均值±标准误（$n=6$）；图柱上的不同字母表示显著差异（$P<0.05$）

芥酸和 IRE1 抑制剂（STF-083010）对草鱼肠细胞中 p-IRE1 蛋白表达的影响如图 5-7 所示。与对照组相比，芥酸会显著上调草鱼肠细胞中 p-IRE1 的蛋白表达（$P<0.05$），而 IRE1 抑制剂（STF-083010）可显著下调由芥酸上调的草鱼肠道细胞中 p-IRE1 的蛋白表达（$P<0.05$）。

芥酸和 IRE1 抑制剂（STF-083010）对草鱼肠细胞中 RhoA 活性的影响如

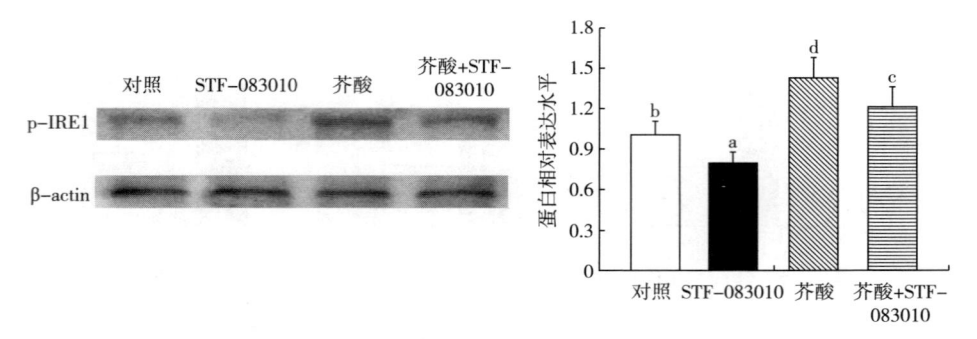

图 5-7 芥酸和 IRE1 抑制剂（STF-083010）处理草鱼肠
细胞后对 p-IRE1 蛋白表达的影响

数据表示为平均值±标准误（$n=6$）；图柱上的不同字母表示显著差异（$P<0.05$）

图 5-8 所示。与对照组相比，芥酸会显著增加草鱼肠细胞中 GTP-RhoA/Total-RhoA 比值（$P<0.05$），而 IRE1 抑制剂（STF-083010）可显著降低由芥酸增加的草鱼肠细胞中 GTP-RhoA/Total-RhoA 比值（$P<0.05$）。

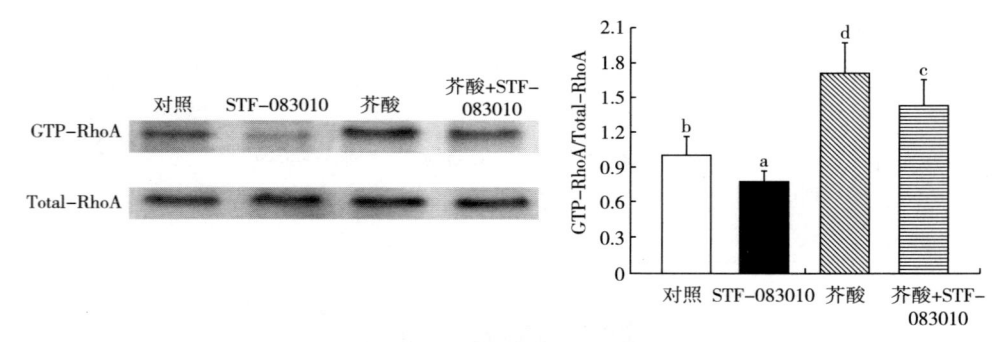

图 5-8 芥酸和 IRE1 抑制剂（STF-083010）处理草鱼肠
细胞后对 GTP-RhoA/Total-RhoA 比值的影响

数据表示为平均值±标准误（$n=6$）；图柱上的不同字母表示显著差异（$P<0.05$）

芥酸对草鱼肠细胞间紧密连接蛋白和黏附连接蛋白相关基因 mRNA 水平的影响如图 5-9 所示。与对照组相比，芥酸下调草鱼肠细胞中紧密连接蛋白 ZO-1、occludin、claudin-7a 和 claudin-7b（$P<0.05$），以及黏附连接蛋白 nectin、E-cadherin 和 β-catenin 的 mRNA 水平（$P<0.05$），同时上调离子通道蛋白 claudin-15a 以及关键信号分子 MLCK、NMII 和 ROCK 的 mRNA 水平（$P<0.05$），而 IRE1 抑制剂（STF-083010）可显著上调由芥酸下调的 ZO-1 和 nectin 的 mRNA 水平（$P<0.05$），同时下调由芥酸上调的 claudin-15a、MLCK、NMII 和 ROCK 的 mRNA 水

平（$P<0.05$）。与此同时，而 IRE1 抑制剂（STF-083010）有上调由芥酸下调的 *occludin*（$P=0.097$）、*claudin-7a*（$P=0.071$）、*claudin-7b*（$P=0.078$）和 *E-cadherin*（$P=0.082$）mRNA 水平的趋势（$P>0.05$）。

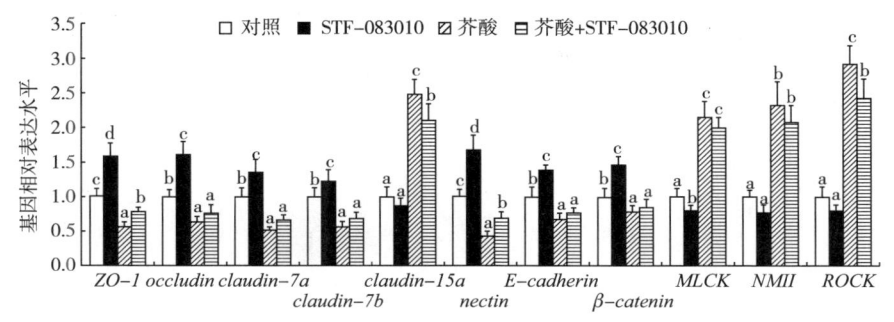

图 5-9 芥酸和 IRE1 抑制剂（STF-083010）处理草鱼肠细胞后对草鱼肠细胞间紧密连接蛋白和黏附连接蛋白相关基因 mRNA 水平的影响

数据表示为平均值±标准误（$n=6$）；图柱上的不同字母表示显著差异（$P<0.05$）

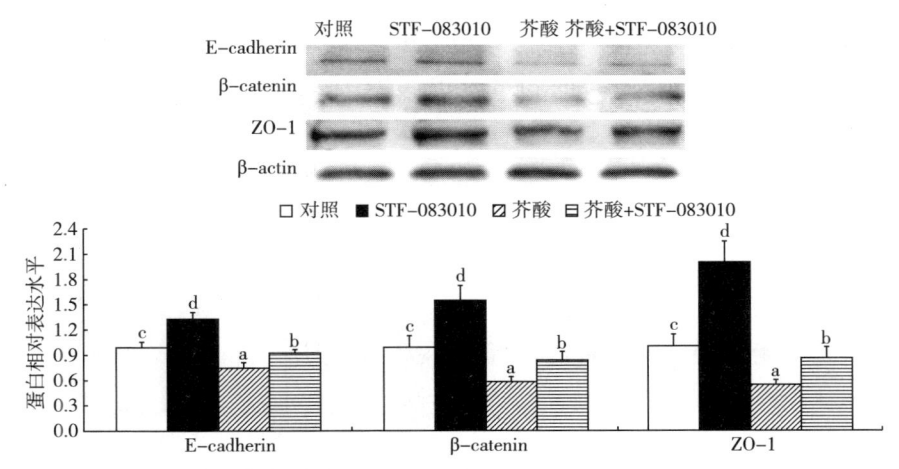

芥酸和 IRE1 抑制剂（STF-083010）对草鱼肠细胞中 E-cadherin、β-catenin 和 ZO-1 蛋白表达的影响如图 5-10 所示。与对照组相比，芥酸可显著下调草鱼肠细胞中 E-cadherin、β-catenin 和 ZO-1 的蛋白表达（$P<0.05$），而 IRE1 抑制剂（STF-083010）可显上调由芥酸下调的草鱼肠道细胞中 E-cadherin、β-catenin 和 ZO-1 的蛋白表达（$P<0.05$）。

图 5-10 芥酸和 IRE1 抑制剂（STF-083010）处理草鱼肠细胞后对 E-cadherin、β-catenin 和 ZO-1 蛋白表达的影响

数据表示为平均值±标准误（$n=6$）；图柱上的不同字母表示显著差异（$P<0.05$）

5.2.3 ATF6信号途径在芥酸破坏草鱼肠细胞间结构完整性中的作用

芥酸和ATF6抑制剂（AEBSF）对草鱼肠细胞中ATF6蛋白表达的影响如图5-11所示。与对照组相比，芥酸会显著上调草鱼肠细胞中ATF6的蛋白表达（$P<0.05$），而ATF6抑制剂（AEBSF）可显著下调由芥酸上调的草鱼肠细胞中ATF6的蛋白表达（$P<0.05$）。

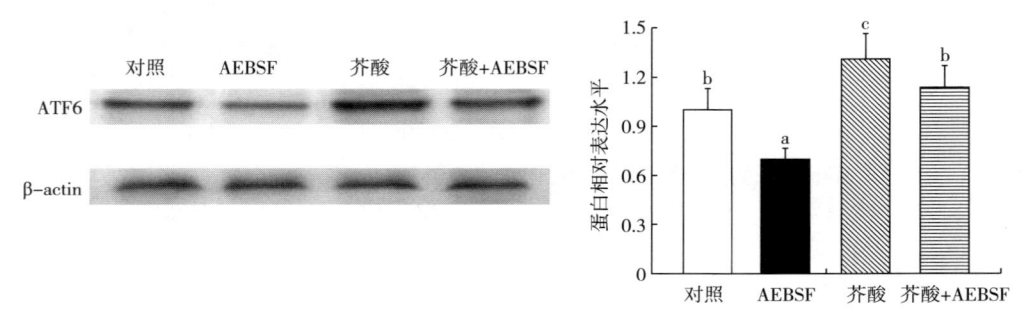

图5-11　芥酸和ATF6抑制剂（AEBSF）处理草鱼肠细胞后对ATF6蛋白表达的影响

数据表示为平均值±标准误（$n=6$）；图柱上的不同字母表示显著差异（$P<0.05$）

芥酸和ATF6抑制剂（AEBSF）对草鱼肠细胞中RhoA活性的影响如图5-12所示。与对照组相比，芥酸可显著增加草鱼肠细胞中GTP-RhoA/Total-RhoA比值（$P<0.05$），同时，ATF6抑制剂（AEBSF）可进一步显著增加由芥酸增加的草鱼肠细胞中GTP-RhoA/Total-RhoA比值（$P<0.05$）。

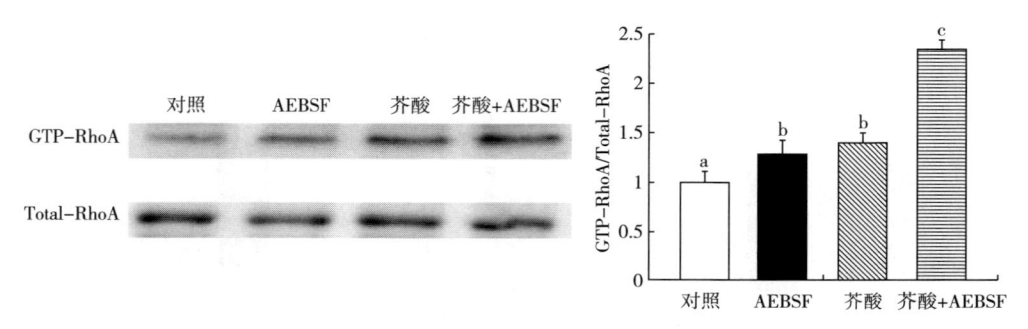

图5-12　芥酸和ATF6抑制剂（AEBSF）处理草鱼肠细胞后对GTP-RhoA/Total-RhoA比值的影响

数据表示为平均值±标准误（$n=6$）；图柱上的不同字母表示显著差异（$P<0.05$）

芥酸对草鱼肠细胞间紧密连接蛋白和黏附连接蛋白相关基因 mRNA 水平的影响如图 5-13 所示。与对照组相比，芥酸下调草鱼肠细胞间紧密连接蛋白 *ZO-1*、*occludin*、*claudin-7a* 和 *claudin-7b*（$P<0.05$），以及黏附连接蛋白 *E-cadherin* 和 *β-catenin* 的 mRNA 水平（$P<0.05$），同时上调离子通道蛋白 *claudin-15a* 以及关键信号分子 *MLCK*、*NMII* 和 *ROCK* 的 mRNA 水平（$P<0.05$），而 ATF6 抑制剂（AEBSF）会进一步下调由芥酸下调的 *ZO-1*、*occludin*、*claudin-7a*、*claudin-7b*、*E-cadherin* 和 *β-catenin* 的 mRNA 水平（$P<0.05$），以及上调由芥酸上调的 *claudin-15a*、*MLCK*、*NMII* 和 *ROCK* 的 mRNA 水平（$P<0.05$）。

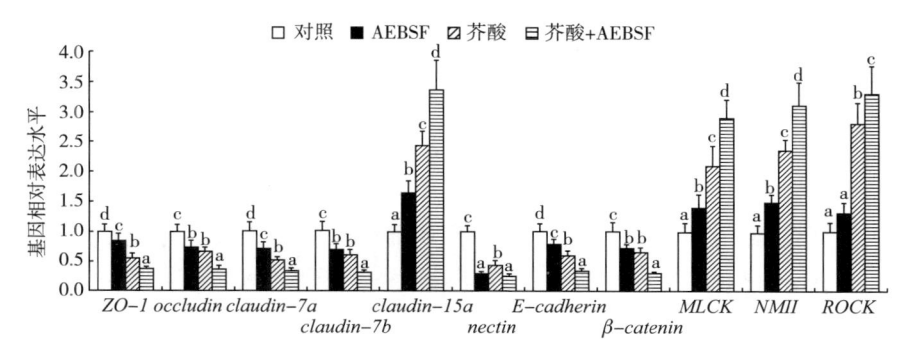

图 5-13 芥酸和 ATF6 抑制剂（AEBSF）处理草鱼肠细胞后对草鱼肠细胞间紧密连接蛋白和黏附连接蛋白相关基因 mRNA 水平的影响

数据表示为平均值±标准误（$n=6$）；图柱上的不同字母表示显著差异（$P<0.05$）

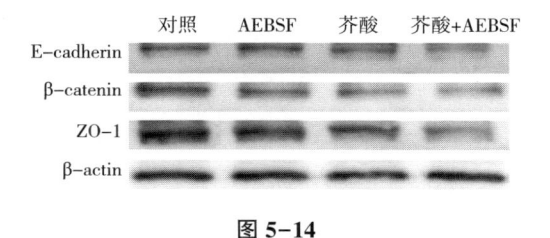

芥酸和 ATF6 抑制剂（AEBSF）对草鱼肠细胞中 E-cadherin、β-catenin 和 ZO-1 蛋白表达的影响如图 5-14 所示。与对照组相比，芥酸可显著下调草鱼肠细胞中 E-cadherin、β-catenin 和 ZO-1 的蛋白表达（$P<0.05$），同时 ATF6 抑制剂（AEBSF）可进一步显著降低由芥酸降低的草鱼肠细胞中 E-cadherin、β-catenin 和 ZO-1 的蛋白表达（$P<0.05$）。

图 5-14

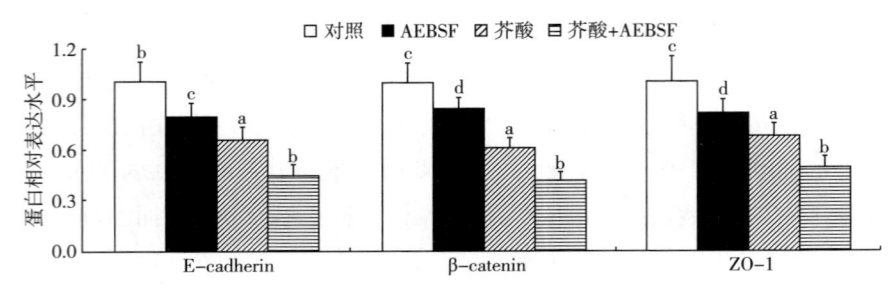

图 5-14　芥酸和 ATF6 抑制剂（AEBSF）处理草鱼肠细胞后对
E-cadherin、β-catenin 和 ZO-1 蛋白表达的影响

数据表示为平均值±标准误（$n=6$）；图柱上的不同字母表示显著差异（$P<0.05$）

5.3　讨论

5.3.1　PERK/eIF2α 信号途径在芥酸破坏草鱼肠细胞间结构完整性中的作用

PERK 是一种内质网 I 型跨膜蛋白，在内质网应激缓解外界刺激的过程中，其会使真核翻译起始因子 2α（eIF2α）失活，从而减少蛋白质翻译。然而，当内质网应激强度增大或持续时间过长，PERK 则会对动物产生一系列的负面影响，比如破坏动物肠上皮屏障结构完整性等。在本研究中，芥酸会下调草鱼肠细胞中 ZO-1、occludin、claudin-7a、claudin-7b、nectin、E-cadherin 和 β-catenin 的基因表达，上调 claudin-15a 的基因表达，和下调 ZO-1、E-cadherin 和 β-catenin 的蛋白表达，以及增加 p-PERK 的蛋白表达和 eIF2α 的基因表达，说明芥酸可能会通过非折叠蛋白反应中的 PERK/eIF2α 信号途径破坏草鱼肠道细胞间结构完整性。GSK2656157 是 PERK 活性特异性抑制剂，被广泛应用于 PERK 生物学功能研究中。在本研究中，与单独添加芥酸相比，在芥酸破坏草鱼肠细胞结构完整性前添加 PERK 活性抑制剂 GSK2656157 可显著降低 p-PERK 的蛋白表达和下调 eIF2α 的基因表达，以及上调 ZO-1、claudin-7a 和 claudin-7b 的基因表达，下调 claudin-15a 的基因表达，和上调 ZO-1、E-cadherin 和 β-catenin 的蛋白表达，这说明芥酸可通过 PERK/eIF2α 信号途径破坏草鱼肠细胞间结构完整性。相似的现象在其他试验结果中也有发现。在鼠上的研究表明：抑制 PERK 会上调脑组织中 ZO-1 和 occludin 的蛋白表达以及睾丸组织中 ZO-1 和 occludin 的蛋白表达。

在人类方面的研究表明：抑制 PERK 会上调静脉内皮细胞中 ZO-1 和 claudin-5 的蛋白表达。与此同时，Fan 等（2017）在大鼠脑中的研究也发现，claudin-5、occludin 和 ZO-1 蛋白表达的升高与 p-PERK 蛋白表达降低有关。以上研究结果说明 PERK 信号分子会参与动物细胞间结构完整性的破坏，并且芥酸可以通过 PERK/eIF2α 信号途径破坏草鱼肠道细胞间结构完整性。除此之外，在本研究中，芥酸会显著增加草鱼肠细胞中 RhoA 的活性，而在芥酸破坏草鱼肠细胞间结构完整性前添加 PERK 活性抑制剂 GSK2656157 可以显著降低 RhoA 的活性。这说明芥酸可通过激活 PERK/eIF2α 信号途径，进而激活 RhoA 信号分子，最终破坏草鱼肠细胞间结构完整性。

5.3.2　IRE1/XBP1 信号途径在芥酸破坏草鱼肠细胞间结构完整性中的作用

IRE1 也是一种内质网Ⅰ型跨膜蛋白，并且是非折叠蛋白反应途径中最保守的信号分子，其主要由激酶和核糖核酸内切酶域构成。当内质网应激发生时，IRE1 具有内切酶活性，其可在 X-盒绑定蛋白 1（XBP1）的 mRNA 上切掉 26 个碱基对片段，从而使无活性的 XBP1 转变成具有活性的 XBP1。具有活性的 XBP1 进入细胞核后会增加伴侣蛋白和其他非折叠蛋白反应相关蛋白的转录，从而增强内质网腔中错误折叠蛋白的降解。然而当内质网应激对动物产生负面影响时，IRE1/XBP1 信号途径也会对动物产生负面影响，如诱导过量的 ROS 产生。在断奶仔猪上的研究表明：p-IRE1 蛋白表达的降低会伴随着其空肠中紧密连接蛋白 ZO-1 和 occludin 蛋白表达的增多。在本研究中，芥酸会下调 *ZO-1*、*occludin*、*claudin-7a*、*claudin-7b*、*nectin*、*E-cadherin* 和 *β-catenin* 的基因表达，上调 *claudin-15a* 的基因表达，和下调 ZO-1、E-cadherin 和 β-catenin 的蛋白表达，以及增加 p-IRE1 的蛋白表达和上调 *XBP1* 的基因表达。这说明芥酸可能也会通过非折叠蛋白反应中的 IRE1/XBP1 信号途径破坏草鱼肠细胞间结构完整性。STF-083010 是 IRE1 的特异性抑制剂，被广泛应用于 IRE1 的生物学功能研究。在本研究中，与单独添加芥酸相比，在芥酸破坏草鱼肠细胞间结构完整性前添加 IRE1 抑制剂 STF-083010 可显著降低 p-IRE1 的蛋白表达和下调 *XBP1* 的基因表达，以及上调 *ZO-1* 和 *nectin* 的基因表达，下调 *claudin-15a* 的基因表达，和上调 ZO-1、E-cadherin 和 β-catenin 的蛋白表达。这说明芥酸也可通过非折叠蛋白反应中的 IRE1/XBP1 信号途径破坏草鱼肠细胞间结构完整性。相似的现象在 Cuevas 等（2017）的研究中也有所报道，其发现 IRE1 的激活会抑制狗肾细胞中 E-cadherin 和 ZO-1 的表达。除此之外，Xie 等（2019）研究表明敲除 IRE1 会降低人结肠癌细胞 HCT116 中 RhoA 的活性。这说明 IRE1 会参与动物细胞间结构完整性的破坏。与此同时，在本研究中，芥酸会显著增加草鱼肠细胞中 RhoA 的活性，而在芥酸破坏草鱼肠细胞间结构完整性前添加 IRE1 活性抑制剂 STF-083010 可以显著降低 RhoA 的活性。这表明芥酸也可通过激活 IRE1/XBP1 信号途径，进而激活 RhoA 信号分子，最终破坏草鱼肠细胞间结构完整性。

5.3.3　ATF6 信号途径在芥酸破坏草鱼肠细胞间结构完整性中的作用

ATF6 是一种内质网 Ⅱ 型跨膜蛋白，该蛋白 N 端胞质区含 CREB/ATF bZIP 结构域，其能调控高尔基复合体中膜内蛋白特异性裂解。当膜蛋白完成裂解后，ATF6 的 N 端胞质区便会移位到细胞核上，最终促进内质网腔中非折叠蛋白的降解和清除。在本研究中，芥酸会下调 *ZO-1*、*occludin*、*claudin-7a*、*claudin-7b*、*nectin*、*E-cadherin* 和 *β-catenin* 的基因表达，上调 *claudin-15a* 的基因表达，和下调 ZO-1、E-cadherin 和 β-catenin 的蛋白表达，以及增加 ATF6 的蛋白表达。这说明芥酸可能也会通过非折叠蛋白反应中的 ATF6 信号途径破坏草鱼肠细胞间结构完整性。AEBSF 是 ATF6 的特异性抑制剂，被广泛应用于 ATF6 的生物学功能研究中。在本研究中，与单独添加芥酸相比，在芥酸破坏草鱼肠细胞间结构完整性前添加 ATF6 抑制剂 AEBSF 会显著降低 ATF6 的蛋白表达，同时进一步显著下调 *ZO-1*、*occludin*、*claudin-7a*、*claudin-7b*、*nectin*、*E-cadherin* 和 *β-catenin* 的基因表达，上调 *claudin-15a* 的基因表达，和下调 ZO-1、E-cadherin 和 β-catenin 的蛋白表达。这说明芥酸不通过非折叠蛋白反应中的 ATF6 信号途径破坏草鱼肠细胞间结构完整性。ATF6 信号途径不参与芥酸破坏草鱼肠细胞间结构完整性可能跟 ATF6 在内质网应激过程中起保护作用有关。研究表明：抑制 ATF6 活性会降低斑马鱼肝细胞的成活率、小鼠的成活率、人神经胶质瘤细胞的增殖和活性以及人黑素瘤细胞的存活。在人类方面的相关研究发现：ATF6 蛋白表达的降低会伴随着人视网膜上皮细胞中紧密连接蛋白 ZO-1 和 occludin 蛋白表达的降低，以及 ATF6 不参与内质网应激破坏人视网膜内皮细胞的屏障完整性这一过程中。这些研究结果说明，ATF6 的激活可能有助于缓解不利因素对动物细胞间结构完整性造成的破坏作用。此外，芥酸会显著增加草鱼肠细胞中的 RhoA 活性，并且在芥酸破坏草鱼肠细胞结构完整性前添加 ATF6 抑制剂 AEBSF 也会显著增加 RhoA 的活性。这说明芥酸不会通过激活 ATF6 信号途径，破坏草鱼肠细胞间结构完整性。综上可知，ATF6 不参与芥酸破坏草鱼肠细胞间结构完整性的过程，这可能跟 ATF6 在芥酸引起草鱼肠细胞内质网应激的过程中起保护草鱼肠道细胞间结构完整性的作用有关。

5.4 小结

根据试验结果，得出以下结论：

（1）芥酸会激活草鱼肠细胞中内质网应激介导的非折叠蛋白反应 PERK/eIF2α、IRE1/XBP1 和 ATF6 信号途径。

（2）芥酸破坏草鱼肠细胞间结构完整性与芥酸激活非折叠蛋白反应 PERK/eIF2α 和 IRE1/XBP1（而不是 ATF6）信号途径，导致信号分子 RhoA 活性增加，进而上调关键信号分子 *MLCK*、*ROCK* 和 *NMII* 的基因表达，下调紧密连接蛋白（*ZO-1*、*occludin*、*claudin-7a* 和 *-7b*）和黏附连接蛋白（*E-cadherin*、*β-catenin* 和 *nectin*）的基因表达，上调离子通道蛋白 *claudin-15a* 的基因表达，以及降低 E-cadherin、β-catenin 和 ZO-1 的蛋白表达有关。

第六章 结论与展望

6.1 结论

本研究首先通过生长试验考察芥酸对生长中期草鱼生产性能、营养物质表观消化率、肠道组织结构、血清二胺氧化酶活性和 D-乳酸含量的影响，探讨芥酸影响生长中期草鱼生产性能的可能因素，并且进一步考察芥酸对生长中期草鱼肠道紧密连接、黏附连接、氧化损伤和凋亡及相关信号分子的影响，研究芥酸影响生长中期草鱼肠道结构的可能机制；其次，通过细胞试验，研究芥酸对草鱼肠细胞关键信号分子 RhoA 的活性、紧密连接和黏附连接相关蛋白的基因和蛋白表达的影响，验证芥酸影响草鱼肠细胞间结构完整性的分子机制；再次，通过生长试验和体外试验，考察芥酸对草鱼肠道细胞超微结构和内质网应激相关蛋白的基因及蛋白表达的影响，研究内质网应激在芥酸影响草鱼肠细胞间结构完整性中的作用及机制；最后，通过细胞试验，研究内质网应激介导的非折叠蛋白反应途径（PERK/eIF2α、IRE1/XBP1 和 ATF6）在芥酸改变草鱼肠细胞间结构完整性中所起的作用。同时根据芥酸对生长中期草鱼生产性能等的影响，确定生长中期草鱼（129.17~471.18 g）饲料中芥酸的控制剂量，为草鱼饲料中菜籽粕的合理添加提供理论基础。

根据本试验的研究结果，可以得出以下结论：

（1）芥酸会降低生长中期草鱼的生产性能和饲料利用效率以及破坏其肠道结构完整性，并且肠道结构完整性的破坏与芥酸破坏草鱼肠道细胞间结构和细胞结构完整性有关。

（2）芥酸破坏草鱼肠细胞间结构完整性与芥酸激活 RhoA 信号途径，进一步上调关键信号分子 *MLCK*、*ROCK* 和 *NMII* 的基因表达，下调紧密连接蛋白和黏附连接蛋白的基因表达，上调离子通道蛋白 *claudin-15a* 的基因表达，以及下调 E-cadherin、β-catenin 和 ZO-1 的蛋白表达有关。

（3）芥酸可诱导草鱼肠道内质网应激，激活 RhoA 信号途径破坏鱼体肠道细胞间结构完整性。

(4) 内质网应激介导的非折叠蛋白反应 PERK/eIF2α 和 IRE1/XBP1（而不是 ATF6）信号途径参与芥酸破坏草鱼肠道细胞间结构完整性的过程中。

(5) 根据增重百分比、前肠 MDA 含量、中肠 ROS 含量以及后肠 PC 含量，确定生长中期草鱼（129.17~471.18 g）饲料中芥酸的控制剂量分别为 0.64%、0.48%、0.48% 和 0.53%。

6.2 创新点

(1) 首次系统揭示了芥酸对鱼类肠道细胞间结构完整性的影响及机制。

(2) 首次考察了芥酸对鱼类肠道内质网应激及其介导的非折叠蛋白反应的影响。

6.3 有待进一步研究的问题

(1) 本实验研究了 Rho 家族中的经典蛋白 RhoA 在芥酸影响草鱼肠道细胞间结构完整性中的作用，但 Rho 家族中的其他经典蛋白，如 Rac 和 Cdc42 蛋白，是否也在芥酸影响鱼类肠道细胞间结构完整性中发挥作用，还有待深入研究。

(2) 本实验研究了芥酸对草鱼肠道内质网应激的影响，有关芥酸引起内质网应激的机制，也有待深入研究。

(3) 本研究发现芥酸可通过诱发非折叠蛋白反应激活 RhoA 信号途径。但非折叠蛋白反应是如何激活 RhoA 信号途径的，该机制也有待进一步研究。

(4) 本研究初步发现芥酸可通过激活 PERK 和 IRE1 信号分子破坏草鱼肠道细胞间结构完整性，然而，芥酸是否介导 PERK 和 IRE1 亚型信号分子参与调控，有待研究。

参考文献

[1] ENAMI H R. A review of using canola/rapeseed meal in aquaculture feeding [J]. Journal of Fisheries and Aquatic Science, 2011, 6 (1): 22-36.

[2] CAI C, SONG L, WANG Y, et al. Assessment of the feasibility of including high levels of rapeseed meal and peanut meal in diets of juvenile crucian carp (*Carassius auratus gibelio* ♀ × *Cyprinus carpio* ♂): Growth, immunity, intestinal morphology, and microflora [J]. Aquaculture, 2013, 410: 203-215.

[3] DOSSOU S, KOSHIO S, ISHIKAWA M, et al. Effect of partial replacement of fish meal by fermented rapeseed meal on growth, immune response and oxidative condition of red sea bream juvenile, *Pagrus major* [J]. Aquaculture, 2018, 490: 228-235.

[4] BU X Y, WANG Y Y, CHEN F Y, et al. An Evaluation of Replacing Fishmeal with Rapeseed Meal in the Diet of *Pseudobagrus ussuriensis*: Growth, Feed Utilization, Nonspecific Immunity, and Growth-related Gene Expression [J]. Journal of the World Aquaculture Society, 2018, 49 (6): 1068-1080.

[5] 张明明, 文华, 蒋明, 等. 饲料菜粕水平对吉富罗非鱼幼鱼生长、肝脏组织结构和部分非特异性免疫指标的影响 [J]. 水产学报, 2011, 35 (5): 748-55.

[6] LUO Y, AI Q, MAI K, et al. Effects of dietary rapeseed meal on growth performance, digestion and protein metabolism in relation to gene expression of juvenile cobia (*Rachycentron canadum*) [J]. Aquaculture, 2012, 368-369: 109-116.

[7] TAN Q, LIU Q, CHEN X, et al. Growth performance, biochemical indices and hepatopancreatic function of grass carp, *Ctenopharyngodon idellus*, would be impaired by dietary rapeseed meal [J]. Aquaculture, 2013, 414: 119-126.

[8] 王永玲. 四种植物蛋白源及其不同添加水平对异育银鲫肠道组织结构的影响 [D]. 苏州: 苏州大学, 2011.

[9] DO S H, KIM B O, FANG L H, et al. Various levels of rapeseed meal in weaning pig diets from weaning to finishing periods [J]. Asian-Australasian journal of

animal sciences, 2017, 30 (9): 1292-1302.

[10] GORRILL A D L, WALKER D M. Rapeseed oils low or high in erucic acid in milk replacers for lambs: Their effects on growth, digestion, nitrogen balance and internal organs [J]. Canadian Journal of Animal Science, 1974, 54 (3): 411-418.

[11] RENNER R, INNIS S M, CLANDININ M T. Effects of high and low erucic acid rapeseed oils on energy metabolism and mitochondrial function of the chick [J]. The Journal of nutrition, 1979, 109 (3): 378-387.

[12] HEISKANEN K M, SAVOLAINEN K M. Erucic acid and erucic acid anilide-induced oxidative burst in human polymorphonuclear leukocytes [J]. Free radical research, 1997, 27 (5): 477-485.

[13] RAMACHANDRAN A, PRABHU R, THOMAS S, et al. Intestinal mucosal alterations in experimental cirrhosis in the rat: role of oxygen free radicals [J]. Hepatology, 2002, 35 (3): 622-629.

[14] DONG Y W, JIANG W D, LIU Y, et al. Threonine deficiency decreased intestinal immunity and aggravated inflammation associated with NF-κB and target of rapamycin signalling pathways in juvenile grass carp (*Ctenopharyngodon idella*) after infection with Aeromonas hydrophila [J]. British Journal of Nutrition, 2017, 118 (2): 92-108.

[15] CHEN K, ZHOU X Q, JIANG W D, et al. Impaired intestinal immune barrier and physical barrier function by phosphorus deficiency: Regulation of TOR, NF-κB, MLCK, JNK and Nrf2 signalling in grass carp (*Ctenopharyngodon idella*) after infection with Aeromonas hydrophila [J]. Fish & shellfish immunology, 2018, 74: 175-189.

[16] P REZ S NCHEZ J, BENEDITO PALOS L, ESTENSORO I, et al. Effects of dietary NEXT ENHANCE© 150 on growth performance and expression of immune and intestinal integrity related genes in gilthead sea bream (*Sparus aurata* L.) [J]. Fish & shellfish immunology, 2015, 44 (1): 117-128.

[17] WEI L, WU P, ZHOU X Q, et al. Dietary silymarin supplementation enhanced growth performance and improved intestinal apical junctional complex on juvenile grass carp (*Ctenopharyngodon idella*) [J]. Aquaculture, 2020: 735311.

[18] WANG Y L, ZHOU X Q, JIANG W D, et al. Effects of dietary zearalenone on

oxidative stress, cell apoptosis, and tight junction in the intestine of juvenile grass carp (*Ctenopharyngodon idella*) [J]. Toxins, 2019, 11 (6): 333.

[19] JIANG W D, HU K, ZHANG J X, et al. Soyabean glycinin depresses intestinal growth and function in juvenile Jian carp (*Cyprinus carpio* var Jian): protective effects of gluta mine [J]. British Journal of Nutrition, 2015, 114 (10): 1569-1583.

[20] 王开卓. 棉酚对草鱼肠道结构和免疫屏障的作用及其机制 [D]. 成都：四川农业大学，2019.

[21] 黎梅. 缩合单宁对生长中期草鱼生产性能、肠道结构和免疫功能的影响及其机制 [D]. 成都：四川农业大学，2019.

[22] FU Q, WANG H, XIA M, et al. The effect of phytic acid on tight junctions in the human intestinal Caco-2 cell line and its mechanism [J]. European Journal of Pharmaceutical Sciences, 2015, 80: 1-8.

[23] DUAN X D, FENG L, JIANG W D, et al. Dietary soybean β-conglycinin suppresses growth performance and inconsistently triggers apoptosis in the intestine of juvenile grass carp (*Ctenopharyngodon idella*) in association with ROS-mediated MAPK signalling [J]. Aquaculture nutrition, 2019, 25 (4): 770-782.

[24] SAUER F D, KRAMER J K G, FORESTER G V, et al. Palmitic and erucic acid metabolism in isolated perfused hearts from weanling pigs [J]. Biochimica et Biophysica Acta (BBA) -Lipids and Lipid Metabolism, 1989, 1004 (2): 205-214.

[25] WEI B, NIE S, MENG Q, et al. Effects of L-carnitine and/or maize distillers dried grains with solubles in diets of gestating and lactating sows on the intestinal barrier functions of their offspring [J]. British Journal of Nutrition, 2016, 116 (3): 459-469.

[26] SATO J, ISHINAGA M, KITO M. Perturbation of arachidonic acid metabolism by erucic acid in rat heart: differential response of the sexes [J]. Agricultural and biological chemistry, 1983, 47 (4): 887-888.

[27] MARTINEZ OROZCO R, NAVARRO TITO N, SOTO GUZMAN A, et al. Arachidonic acid promotes epithelial to mesenchymal like transition in mammary epithelial cells MCF10A [J]. European journal of cell biology, 2010, 89 (6): 476-488.

[28] K LLER M, WACHTLER P, DAVID A, et al. Arachidonic acid induces DNA-fragmentation in human polymorphonuclear neutrophil granulocytes [J]. Inflammation, 1997, 21 (5): 463-474.

[29] VEMURI A K, ACHARYA V, PONDAY L R, et al. Transgenic zero-erucic and high-oleic mustard oil improves glucose clearance rate, erythrocyte membrane docosahexaenoic acid content and reduces osmotic fragility of erythrocytes in male Syrian golden hamsters [J]. Journal of nutrition & intermediary metabolism, 2018, 12: 28-35.

[30] ZHANG Y, PENG F, GAO B, et al. High glucose-induced RhoA activation requires caveolae and PKCβ1-mediated ROS generation [J]. American Journal of Physiology-Renal Physiology, 2011, 302 (1): F159-F172.

[31] STAM H, GEELHOED MIERAS T, H LSMANN W C. Erucic acid-induced alteration of cardiac triglyceride hydrolysis [J]. Lipids, 1980, 15 (4): 242-250.

[32] FENG L, NI P J, JIANG W D, et al. Decreased enteritis resistance ability by dietary low or excess levels of lipids through impairing the intestinal physical and immune barriers function of young grass carp (*Ctenopharyngodon idella*) [J]. Fish & shellfish immunology, 2017, 67: 493-512.

[33] MURPHY C C, MURPHY E J, GOLOVKO M Y. Erucic acid is differentially taken up and metabolized in rat liver and heart [J]. Lipids, 2008, 43 (5): 391-400.

[34] STEIN D T, STEVENSON B E, CHESTER M W, et al. The insulinotropic potency of fatty acids is influenced profoundly by their chain length and degree of saturation [J]. The Journal of clinical investigation, 1997, 100 (2): 398-403.

[35] GHOSH A, ABDO S, ZHAO S L, et al. Insulin Inhibits Nrf2 Gene Expression via Heterogeneous Nuclear Ribonucleoprotein F/K in Diabetic Mice [J]. Endocrinology, 2017, 158 (4): 903-919.

[36] THOMASSEN M S, HELGERUD P, NORUM K R. Chain-shortening of erucic acid and microperoxisomal β-oxidation in rat small intestine [J]. Biochemical journal, 1985, 225 (2): 301-306.

[37] LAGER S, GACCIOLI F, RAMIREZ V I, et al. Oleic acid stimulates system A

amino acid transport in primary human trophoblast cells mediated by toll-like receptor 4 [J]. Journal of lipid research, 2013, 54 (3): 725-733.

[38] WU X, SUN L, ZHA W, et al. HIV protease inhibitors induce endoplasmic reticulum stress and disrupt barrier integrity in intestinal epithelial cells [J]. Gastroenterology, 2010, 138 (1): 197-209.

[39] DIGALEH H, KIAEI M, KHODAGHOLI F. Nrf2 and Nrf1 signaling and ER stress crosstalk: implication for proteasomal degradation and autophagy [J]. Cellular and molecular life sciences, 2013, 70 (24): 4681-4694.

[40] SONG Y F, LUO Z, ZHANG L H, et al. Endoplasmic reticulum stress and disturbed calcium homeostasis are involved in copper-induced alteration in hepatic lipid metabolism in yellow catfish *Pelteobagrus fulvidraco* [J]. Chemosphere, 2016, 144: 2443-2453.

[41] LI J, XIA X, KE Y, et al. Trichosanthin induced apoptosis in HL-60 cells via mitochondrial and endoplasmic reticulum stress signaling pathways [J]. Biochimica et Biophysica Acta (BBA) -General Subjects, 2007, 1770 (8): 1169-1180.

[42] BIAN F, JIANG H, MAN M, et al. Dietary gossypol suppressed postprandial TOR signaling and elevated ER stress pathways in turbot (*Scophthalmus maximus* L.) [J]. American Journal of Physiology - Endocrinology and Metabolism, 2016, 312 (1): E37-E47.

[43] NAGESH P, HATAMI E, CHOWDHURY P, et al. Tannic acid induces endoplasmic reticulum stress-mediated apoptosis in prostate Cancer [J]. Cancers, 2018, 10 (3): 68.

[44] PASINI E, CARGNONI A, CONDORELLI E, et al. Effect of prolonged treatment with propionyl-L-carnitine on erucic acid-induced myocardial dysfunction in rats [J]. Molecular and cellular biochemistry, 1992, 112 (2): 117-123.

[45] KHARROUBI I, LADRI RE L, CARDOZO A K, et al. Free fatty acids and cytokines induce pancreatic β-cell apoptosis by different mechanisms: role of nuclear factor-κB and endoplasmic reticulum stress [J]. Endocrinology, 2004, 145 (11): 5087-5096.

[46] FAN L, HU L, YANG B, et al. Erlotinib promotes endoplasmic reticulum stress-mediated injury in the intestinal epithelium [J]. Toxicology and applied pharma-

cology, 2014, 278 (1): 45-52.

[47] HE Z, ZHOU Y, WANG Q, et al. Inhibiting endoplasmic reticulum stress by lithium chloride contributes to the integrity of blood-spinal cord barrier and functional recovery after spinal cord injury [J]. American journal of translational research, 2017, 9 (3): 1012.

[48] YUAN X, WANG J, LI Y, et al. Allergy immunotherapy restores airway epithelial barrier dysfunction through suppressing IL-25-induced endoplasmic reticulum stress in asthma [J]. Scientific reports, 2018, 8 (1): 1-11.

[49] FAO. Fisheries and Aquaculture Statistics [J]. Food and Agriculture Organization of United Nations, 2018.

[50] NRC. Nutrient requirements of fish and shrimp [M]. The National Academies Press Washington, DC, USA. 2011.

[51] FRIEND D W, GILKA F, CORNER A H. Growth, carcass quality and cardiopathology of boars and gilts fed diets containing rapeseed and soybean oils [J]. Canadian Journal of Animal Science, 1975, 55 (4): 571-578.

[52] ZHANG L S, TAN Y, OUYANG Y L, et al. Effects of high erucic acid rapeseed oil on fatty acid oxidation in rat liver [J]. Biomedical and environmental sciences: BES, 1991, 4 (3): 262-267.

[53] ASTORG P O. Heart lipidosis induced by short-term feeding of cis-or trans-docosenoic acids in weanling or 7-week-old rats [J]. Annals of Nutrition and Metabolism, 1981, 25 (4): 201-207.

[54] KRAMER J K G, HULAN H W. Changes in the acyl and alkenyl group composition of cardiac phospholipids in boars fed corn oil or rapeseed oil [J]. Lipids, 1977, 12 (2): 159-164.

[55] AKIBA Y, MATSUMOTO T. Effects of graded doses of goitrin, a goitrogen in rapeseed, on synthesis and release of thyroid hormone in chicks [J]. Nippon Chikusan Gakkai-Ho, 1977, 48 (12): 757-765.

[56] BOURDON D, AUMAITRE A. Low-glucosinolate rapeseeds and rapeseed meals: effect of technological treatments on chemical composition, digestible energy content and feeding value for growing pigs [J]. Animal Feed Science and Technology, 1990, 30 (3-4): 175-191.

[57] LO M T, HILL D C. Effect of feeding a high level of rapeseed meal on weight

gains and thyroid function of rats [J]. The Journal of nutrition, 1971, 101 (8): 975-980.

[58] VIRTANEN A I, KREULA M, KIESVAARA M. The transfer of L-5-vinyl-2-thio-oxazolidone (oxazolidone-thione) to milk [J]. Acta Chemica Scandinavica, 1958, 12: 580-581.

[59] CHEN S, ANDREASSON E. Update on glucosinolate metabolism and transport [J]. Plant Physiology and Biochemistry, 2001, 39 (9): 743-758.

[60] TRIPATHI M K, MISHRA A S. Glucosinolates in animal nutrition: A review [J]. Animal Feed Science and Technology, 2007, 132 (1): 1-27.

[61] SKUGOR S, JODAA HOLM H, BJELLAND A K, et al. Nutrigenomic effects of glucosinolates on liver, muscle and distal kidney in parasite-free and salmon louse infected Atlantic salmon [J]. Parasites & vectors, 2016, 9 (1): 639.

[62] BUREL C, BOUJARD T, ESCAFFRE A M, et al. Dietary low-glucosinolate rapeseed meal affects thyroid status and nutrient utilization in rainbow trout (*Oncorhynchus mykiss*) [J]. British Journal of Nutrition, 2000, 83 (6): 653-664.

[63] VON DANWITZ A, SCHULZ C. Effects of dietary rapeseed glucosinolates, sinapic acid and phytic acid on feed intake, growth performance and fish health in turbot (*Psetta maxima* L.) [J]. Aquaculture, 2020, 516: 734624.

[64] VUORELA S, SALMINEN H, M KEL M, et al. Effect of plant phenolics on protein and lipid oxidation in cooked pork meat patties [J]. Journal of agricultural and food chemistry, 2005, 53 (22): 8492-8497.

[65] NACZK M, AMAROWICZ R, SULLIVAN A, et al. Current research developments on polyphenolics of rapeseed/canola: a review [J]. Food Chemistry, 1998, 62 (4): 489-502.

[66] PAPATRYPHON E, HOWELL R A, SOARES JR J H. Growth and mineral absorption by striped bass *Morone saxatilis* fed a plant feedstuff based diet supplemented with phytase [J]. Journal of the World Aquaculture Society, 1999, 30 (2): 161-173.

[67] SPINELLI J, HOULE C R, WEKELL J C. The effect of phytates on the growth of rainbow trout (*Salmo gairdneri*) fed purified diets containing varying quantities of calcium and magnesium [J]. Aquaculture, 1983, 30 (1-4): 71-83.

[68] ZHONG J R, FENG L, JIANG W D, et al. Phytic acid disrupted intestinal im-

mune status and suppressed growth performance in on-growing grass carp (*Ctenopharyngodon idella*) [J]. Fish & shellfish immunology, 2019, 92: 536-551.

[69] LIU L, LIANG X F, LI J, et al. Feed intake, feed utilization and feeding-related gene expression response to dietary phytic acid for juvenile grass carp (*Ctenopharyngodon idellus*) [J]. Aquaculture, 2014, 424-425: 201-206.

[70] LIU L W, LIANG X F, LI J, et al. Effects of supplemental phytic acid on the apparent digestibility and utilization of dietary amino acids and minerals in juvenile grass carp (*Ctenopharyngodon idellus*) [J]. Aquaculture nutrition, 2018, 24 (2): 850-857.

[71] LAINING A, TRAIFALGAR R, THU M, et al. Influence of Dietary Phytic Acid on Growth, Feed Intake, and Nutrient Utilization in Juvenile Japanese Flounder, *Paralichthys olivaceus* [J]. Journal of the World Aquaculture Society, 2010, 41 (5): 746-755.

[72] CHOWDHURY M A K, MARTIE T, BUREAU D. Effect of dietary phytic acid and semi-purified lignin on energy storage indices, growth performance, nutrient and energy partitioning of rainbow trout, *Oncorhynchus mykiss* [J]. Aquaculture nutrition, 2015, 21 (6): 843-852.

[73] KHAN A, GHOSH K. Phytic acid-induced inhibition of digestive protease and α-amylase in three Indian major carps: An in vitro study [J]. Journal of the World Aquaculture Society, 2013, 44 (6): 853-859.

[74] CHUNG K T, WEI C I, JOHNSON M G. Are tannins a double-edged sword in biology and health? [J]. Trends in Food Science & Technology, 1998, 9 (4): 168-175.

[75] BUYUKCAPAR H M, ATALAY A, KAMALAK A. Growth performance of Nile tilapia (Oreochromis niloticus) fed with diets containing different levels of hydrolysable and condensed tannin [J]. Journal of Agricultural Science and Technology, 2011, 13 (7): 1045-1051.

[76] LI M, FENG L, JIANG W D, et al. Condensed tannins decreased the growth performance and impaired intestinal immune function in on-growing grass carp (*Ctenopharyngodon idella*) [J]. British Journal of Nutrition, 2019, 1-41.

[77] YAO J, CHEN P, APRAKU A, et al. Hydrolysable Tannin Supplementation

Alters Digestibility and Utilization of Dietary Protein, Lipid, and Carbohydrate in Grass Carp (*Ctenopharyngodon idellus*) [J]. Frontiers in Nutrition, 2019, 6: 183.

[78] OMNES M H, LE GOASDUFF J, LE DELLIOU H, et al. Effects of dietary tannin on growth, feed utilization and digestibility, and carcass composition in juvenile European seabass (*Dicentrarchus labrax* L.) [J]. Aquaculture Reports, 2017, 6: 21-27.

[79] TALUKDAR S, GHOSH K. Differential inhibition of digestive proteases by tannin in two size groups of rohu (*Labeo rohita*, Hamilton): A biochemical and zymography study [J]. Aquaculture research, 2019, 50 (2): 449-456.

[80] SIM J S, TOY B, CRICK D C, et al. Effect of dietary erucic acid on the utilization of oils or fats by growing chicks [J]. Poultry science, 1985, 64 (11): 2150-2154.

[81] 吴关庭, 郎春秀, 陈锦清. 芥酸的生产及其衍生产品开发 [J]. 中国油脂, 32 (6): 27-31.

[82] CHAIN E P O C I T F, KNUTSEN H K, ALEXANDER J, et al. Erucic acid in feed and food [J]. EFSA Journal, 2016, 14 (11): e04593.

[83] LEE I K, KYE Y C, KIM G, et al. Stress, nutrition, and intestinal immune responses in pigs—a review [J]. Asian-Australasian journal of animal sciences, 2016, 29 (8): 1075-1082.

[84] EATON S, BARTLETT K B, POURFARZAM M. Mammalian mitochondrial β-oxidation [J]. Biochemical journal, 1996, 320 (2): 345-357.

[85] CHRISTOPHERSEN B O, BREMER J. Erucic acid—an inhibitor of fatty acid oxidation in the heart [J]. Biochimica et Biophysica Acta (BBA) -Lipids and Lipid Metabolism, 1972, 280 (4): 506-514.

[86] BARTLETT K, EATON S. Mitochondrial β-oxidation [J]. European Journal of Biochemistry, 2004, 271 (3): 462-469.

[87] BOENGLER K, KOSIOL M, MAYR M, et al. Mitochondria and ageing: role in heart, skeletal muscle and adipose tissue [J]. Journal of Cachexia, Sarcopenia and Muscle, 2017, 8 (3): 349-369.

[88] SCHIEFER B, LOEW F M, LAXDAL V, et al. Morphologic effects of dietary plant and animal lipids rich in docosenoic acids on heart and skeletal muscle of

cynomolgus monkeys [J]. The American journal of pathology, 1978, 90 (3): 551.

[89] GOPALAN C, KRISHNAMURTHI D, SHENOLIKAR I S, et al. Myocardial changes in monkeys fed mustard oil [J]. Annals of Nutrition and Metabolism, 1974, 16 (6): 352-365.

[90] LEE S H C, CLANDININ M T. Effect of dietary fat on the utilization of fatty acids by myocardial tissue in the rat [J]. The Journal of nutrition, 1986, 116 (11): 2096-2105.

[91] FRIESEN D L, SINGH A. Liver ultrastructure in pigs fed various oils [J]. Research in veterinary science, 1981, 30 (3): 368-373.

[92] HULAN H W, HUNSAKER W G, KRAMER J K G, et al. The development of dermal lesions and alopecia in male rats fed rapeseed oil [J]. Canadian journal of physiology and pharmacology, 1976, 54 (1): 1-6.

[93] PINSON A, PADIEU P. Erucic acid oxidation by beating heart cells in culture [J]. FEBS letters, 1974, 39 (1): 88-90.

[94] ROGERS C G. Erucic acid and phospholipids of newborn rat heart cells in culture [J]. Lipids, 1977, 12 (4): 375-381.

[95] CHRISTENSEN E, HAGVE T-A, CHRISTOPHERSEN B O. The Zellweger syndrome: deficient chain-shortening of erucic acid [22: 1 ($n-9$)] and adrenic acid [22: 4 ($n-6$)] in cultured skin fibroblasts [J]. Biochimica et Biophysica Acta (BBA) -Lipids and Lipid Metabolism, 1988, 959 (2): 95-99.

[96] NORSETH J, CHRISTOPHERSEN B O. Chain shortening of erucic acid in isolated liver cells [J]. FEBS letters, 1978, 88 (2): 353-357.

[97] CHRISTIANSEN E N, GRAY T J B, LAKE B G. Effect of very long chain fatty acids on peroxisomal β-oxidation in primary rat hepatocyte cultures [J]. Lipids, 1985, 20 (12): 929-932.

[98] VAN HERPEN N A, SCHRAUWEN-HINDERLING V B. Lipid accumulation in non-adipose tissue and lipotoxicity [J]. Physiology & behavior, 2008, 94 (2): 231-241.

[99] KIM Y S, HO S B. Intestinal goblet cells and mucins in health and disease: recent insights and progress [J]. Current gastroenterology reports, 2010, 12 (5): 319-330.

[100] FEDERICI G, SHAW B J, HANDY R D. Toxicity of titanium dioxide nanoparticles to rainbow trout (*Oncorhynchus mykiss*): Gill injury, oxidative stress, and other physiological effects [J]. Aquatic toxicology, 2007, 84 (4): 415-430.

[101] YU L C H, FLYNN A N, TURNER J R, et al. SGLT-1-mediated glucose uptake protects intestinal epithelial cells against LPS-induced apoptosis and barrier defects: a novel cellular rescue mechanism? [J]. FASEB journal : official publication of the Federation of American Societies for Experimental Biology, 2005, 19 (13): 1822-1835.

[102] TSIKAS D. Assessment of lipid peroxidation by measuring malondialdehyde (MDA) and relatives in biological samples: Analytical and biological challenges [J]. Analytical Biochemistry, 2017, 524: 13-30.

[103] CHEVION M, BERENSHTEIN E, STADTMAN E R. Human studies related to protein oxidation: protein carbonyl content as a marker of damage [J]. Free radical research, 2000, 33 Suppl: S99-108.

[104] CHEN M, YIN J, LIANG Y, et al. Oxidative stress and immunotoxicity induced by graphene oxide in zebrafish [J]. Aquatic toxicology, 2016, 174: 54-60.

[105] HUANG C, WU P, JIANG W D, et al. Deoxynivalenol decreased the growth performance and impaired intestinal physical barrier in juvenile grass carp (*Ctenopharyngodon idella*) [J]. Fish & shellfish immunology, 2018, 80: 376-391.

[106] JIANG W D, ZHOU X Q, ZHANG L, et al. Vita min A deficiency impairs intestinal physical barrier function of fish [J]. Fish & shellfish immunology, 2019, 87: 546-558.

[107] NAKAJIMA H, NAKAJIMA-TAKAGI Y, TSUJITA T, et al. Tissue-Restricted Expression of Nrf2 and Its Target Genes in Zebrafish with Gene-Specific Variations in the Induction Profiles [J]. PloS one, 2011, 6 (10): e26884.

[108] BORTNER C D, OLDENBURG N B E, CIDLOWSKI J A. The role of DNA fragmentation in apoptosis [J]. Trends in Cell Biology, 1995, 5 (1): 21-26.

[109] SHI Y. Mechanisms of Caspase Activation and Inhibition during Apoptosis [J].

Molecular Cell, 2002, 9 (3): 459-470.

[110] KRUMSCHNABEL G, PODRABSKY J E. Fish as model systems for the study of vertebrate apoptosis [J]. Apoptosis, 2009, 14 (1): 1-21.

[111] CORY S, ADAMS J M. The Bcl2 family: regulators of the cellular life-or-death switch [J]. Nature Reviews Cancer, 2002, 2 (9): 647-656.

[112] MIRKES P. 2001 Warkany lecture: To die or not to die, the role of apoptosis in normal and abnormal mammalian development [J]. Teratology, 2002, 65: 228-239.

[113] CHOI S Y, KIM M J, KANG C M, et al. Activation of Bak and Bax through c-Abl-Protein Kinase Cδ-p38 MAPK Signaling in Response to Ionizing Radiation in Human Non-small Cell Lung Cancer Cells [J]. J. biol. chem, 2006, 281 (11): 7049-7059.

[114] DE CHIARA G, MARCOCCI M E, TORCIA M, et al. Bcl-2 Phosphorylation by p38 MAPK: Identification of target sites and biologic consequences [J]. The Journal of biological chemistry, 2006, 281: 21353-21361.

[115] SU J L, LIN M T, HONG C C, et al. Resveratrol induces FasL-related apotosis through Cdc42 activation of ASK1/JNK-dependent signaling pathway in human leukemia HL-60 cells [J]. Carcinogenesis, 2005, 26: 1-10.

[116] PAPADAKIS E S, FINEGAN K G, WANG X, et al. The regulation of Bax by c-Jun N-ter minal protein kinase (JNK) is a prerequisite to the mitochondrial-induced apoptotic pathway [J]. FEBS letters, 2006, 580 (5): 1320-1326.

[117] HOUTSMULLER U M T, STRUIJK C B, VAN DER BEEK A. Decrease in rate of ATP synthesis of Isolated rat heart mitochondria induced by dietary erucic acid [J]. Biochimica et Biophysica Acta (BBA) - Lipids and Lipid Metabolism, 1970, 218 (3): 564-566.

[118] CHEN Y P, JIANG W D, LIU Y, et al. Exogenous phospholipids supplementation improves growth and modulates immune response and physical barrier referring to NF-κB, TOR, MLCK and Nrf2 signaling factors in the intestine of juvenile grass carp (*Ctenopharyngodon idella*) [J]. Fish & shellfish immunology, 2015, 47 (1): 46-62.

[119] IVANOV A I. Actin motors that drive formation and disassembly of epithelial apical junctions [J]. Front Biosci, 2008, 13 (500): 6662-6681.

[120] ZIHNI C, MILLS C, MATTER K, et al. Tight junctions: from simple barriers to multifunctional molecular gates [J]. Nat Rev Mol Cell Biol, 2016, 17 (9): 564-580.

[121] GONZ LEZ MARISCAL L, BETANZOS A, NAVA P, et al. Tight junction proteins [J]. Progress in Biophysics & Molecular Biology, 2003, 81 (1): 1-44.

[122] CAMPBELL H K, MAIERS J L, DEMALI K A. Interplay between tight junctions & adherens junctions [J]. Experimental Cell Research, 2017, 358 (1): 39-44.

[123] QUIROS M, NUSRAT A. RhoGTPases, actomyosin signaling and regulation of the epithelial Apical Junctional Complex [C]. Se minars in cell & developmental biology, 2014: 194-203.

[124] GUSTI V, BENNETT K M, LO D D. CD137 signaling enhances tight junction resistance in intestinal epithelial cells [J]. Physiological Reports, 2014, 2 (8): e12090.

[125] ULLUWISHEWA D, ANDERSON R C, MCNABB W C, et al. Regulation of Tight Junction Permeability by Intestinal Bacteria and Dietary Components [J]. Journal of Nutrition, 2011, 141 (5): 769-776.

[126] HSU K L, FAN H J, CHEN Y C, et al. Protein kinase C-Fyn kinase cascade mediates the oleic acid-induced disassembly of neonatal rat cardiomyocyte adherens junctions [J]. The international journal of biochemistry & cell biology, 2009, 41 (7): 1536-1546.

[127] SHAO P, ZHU J, DING H, et al. Tripterygium hypoglaucum (Levl.) Hutch attenuates oleic acid-induced acute lung injury in rats through up-regulating claudin-5 and ZO-1 expression [J]. International Journal of Clinical and Experimental Medicine, 2018, 11: 6634-6647.

[128] TERRY S, NIE M, MATTER K, et al. Rho Signaling and Tight Junction Functions [J]. Physiology, 2010, 25 (1): 16-26.

[129] WENNERBERG K, ROSSMAN K L, DER C J. The Ras superfamily at a glance [J]. Journal of cell science, 2005, 118 (5): 843-846.

[130] HALL A. Rho family gtpases. Portland Press Ltd., 2012.

[131] NUSRAT A, GIRY M, TURNER J R, et al. Rho protein regulates tight junc-

tions and perijunctional actin organization in polarized epithelia [J]. Proceedings of the National Academy of Sciences, 1995, 92 (23): 10629-10633.

[132] JOU T S, SCHNEEBERGER E E, JAMES NELSON W. Structural and functional regulation of tight junctions by RhoA and Rac1 small GTPases [J]. The Journal of cell biology, 1998, 142 (1): 101-115.

[133] BRUEWER M, HOPKINS A M, HOBERT M E, et al. RhoA, Rac1, and Cdc42 exert distinct effects on epithelial barrier via selective structural and biochemical modulation of junctional proteins and F-actin [J]. American Journal of Physiology-Cell Physiology, 2004, 287 (2): C327-C335.

[134] HOPKINS A M, WALSH S V, VERKADE P, et al. Constitutive activation of Rho proteins by CNF-1 influences tight junction structure and epithelial barrier function [J]. Journal of cell science, 2003, 116 (4): 725-742.

[135] HUANG N, ZHANG X, JIANG Y, et al. Increased levels of serum pigment epithelium-derived factor aggravate proteinuria via induction of podocyte actin rearrangement [J]. International urology and nephrology, 2019, 51 (2): 359-367.

[136] FENG S, ZOU L, WANG H, et al. RhoA/ROCK-2 pathway inhibition and tight junction protein upregulation by catalpol suppresses lipopolysaccharide-induced disruption of blood-brain barrier permeability [J]. Molecules, 2018, 23 (9): 2371.

[137] CITAL N MADRID A F, VARGAS-ROBLES H, GARC A-PONCE A, et al. Cortactin deficiency causes increased RhoA/ROCK1-dependent actomyosin contractility, intestinal epithelial barrier dysfunction, and disproportionately severe DSS-induced colitis [J]. Mucosal Immunology, 2017, 10 (5): 1237-1247.

[138] LARYEA M D, JIANG Y F, XU G L, et al. Fatty acid composition of blood lipids in Chinese children consu ming high erucic acid rapeseed oil [J]. Annals of Nutrition and Metabolism, 1992, 36 (5-6): 273-278.

[139] CHEN C J, XIAO P, CHEN Y, et al. Selenium Deficiency Affects Uterine Smooth Muscle Contraction Through Regulation of the RhoA/ROCK Signalling Pathway in Mice [J]. Biological trace element research, 2019, 1-10.

[140] CHEVET E, CAMERON P H, PELLETIER M F, et al. The endoplasmic re-

ticulum: integration of protein folding, quality control, signaling and degradation [J]. Current Opinion in Structural Biology, 2001, 11 (1): 120-124.

[141] WU R F, MA Z, LIU Z, et al. Nox4-Derived H_2O_2 Mediates Endoplasmic Reticulum Signaling through Local Ras Activation [J]. Molecular and cellular biology, 2010, 30 (14): 3553.

[142] RON D, WALTER P. Signal integration in the endoplasmic reticulum unfolded protein response [J]. Nature reviews Molecular cell biology, 2007, 8 (7): 519-529.

[143] RUTKOWSKI D T, HEGDE R S. Regulation of basal cellular physiology by the homeostatic unfolded protein response [J]. The Journal of cell biology, 2010, 189 (5): 783-794.

[144] TSANG K Y, CHAN D, BATEMAN J F, et al. In vivo cellular adaptation to ER stress: survival strategies with double-edged consequences [J]. Journal of cell science, 2010, 123 (13): 2145-2154.

[145] HOTAMISLIGIL G S. Endoplasmic reticulum stress and the inflammatory basis of metabolic disease [J]. Cell, 2010, 140 (6): 900-917.

[146] BERTOLOTTI A, ZHANG Y, HENDERSHOT L M, et al. Dynamic interaction of BiP and ER stress transducers in the unfolded-protein response [J]. Nature cell biology, 2000, 2 (6): 326-332.

[147] SHEN J, CHEN X, HENDERSHOT L, et al. ER stress regulation of ATF6 localization by dissociation of BiP/GRP78 binding and unmasking of Golgi localization signals [J]. Developmental cell, 2002, 3 (1): 99-111.

[148] HETZ C. The unfolded protein response: controlling cell fate decisions under ER stress and beyond [J]. Nature reviews Molecular cell biology, 2012, 13 (2): 89-102.

[149] KITAMURA M. The unfolded protein response triggered by environmental factors [C]. Se minars in immunopathology, 2013: 259-275.

[150] ZHANG G, WANG Z, CHEN W, et al. Dual effects of gossypol on human hepatocellular carcinoma via endoplasmic reticulum stress and autophagy [J]. The international journal of biochemistry & cell biology, 2019, 113: 48-57.

[151] SODERQUIST R S, DANILOV A V, EASTMAN A. Gossypol Increases Expression of the Pro-apoptotic BH_3-only Protein NOXA through a Novel Mecha-

nism Involving Phospholipase A2, Cytoplasmic Calcium, and Endoplasmic Reticulum Stress [J]. Journal of biological chemistry, 2014, 289 (23): 16190-16199.

[152] TAGAWA Y, HIRAMATSU N, KASAI A, et al. Induction of apoptosis by cigarette smoke via ROS-dependent endoplasmic reticulum stress and CCAAT/enhancer-binding protein-homologous protein (CHOP) [J]. Free Radical Biology and Medicine, 2008, 45 (1): 50-59.

[153] HUANG Y, WANG Y, FENG Y, et al. Role of endoplasmic reticulum stress-autophagy axis in severe burn-induced intestinal tight junction barrier dysfunction in mice [J]. Frontiers in physiology, 2019, 10: 606.

[154] LIANG B, WANG S, WANG Q, et al. Aberrant endoplasmic reticulum stress in vascular smooth muscle increases vascular contractility and blood pressure in mice deficient of AMP-activated protein kinase-α2 in vivo [J]. Arteriosclerosis, thrombosis, and vascular biology, 2013, 33 (3): 595-604.

[155] BOUCHECAREILH M, MARZA E, CARUSO M E, et al. Small GTPase Signaling and the Unfolded Protein Response [J]. Methods in Enzymology, 2011, 491: 343-360.

[156] SEO S H, KIM S E, LEE S E. ER stress induced by ER calcium depletion and UVB irradiation regulates tight junction barrier integrity in human keratinocytes [J]. Journal of Dermatological Science, 2020, 98 (1): 41-49.

[157] TSUTSUMI S, GOTOH T, TOMISATO W, et al. Endoplasmic reticulum stress response is involved in nonsteroidal anti-inflammatory drug-induced apoptosis [J]. Cell death and differentiation, 2004, 11 (9): 1009-1016.

[158] TORNIN J, HERMIDA-PRADO F, PADDA R S, et al. FUS-CHOP promotes invasion in myxoid liposarcoma through a SRC/FAK/RHO/ROCK-dependent pathway [J]. Neoplasia, 2018, 20 (1): 44-56.

[159] ZHAI J, LIN H, NIE Z, et al. Direct interaction of focal adhesion kinase with p190RhoGEF [J]. Journal of biological chemistry, 2003, 278 (27): 24865-24873.

[160] XIE Y, LIU C, QIN Y, et al. Knockdown of IRE1α suppresses metastatic potential of colon cancer cells through inhibiting FN1-Src/FAK-GTPases signaling [J]. The international journal of biochemistry & cell biology, 2019,

114：105572.

[161] BELL J M. Nutrients and toxicants in rapeseed meal: a review [J]. Journal of animal science, 1984, 58 (4): 996-1010.

[162] LI X Y, TANG L, HU K, et al. Effect of dietary lysine on growth, intestinal enzymes activities and antioxidant status of sub-adult grass carp (*Ctenopharyngodon idella*) [J]. Fish physiology and biochemistry, 2014, 40 (3): 659-671.

[163] 段绪东. β-伴大豆球蛋白对草鱼肠道细胞凋亡的作用与机制研究 [D]. 成都：四川农业大学，2019.

[164] KRAMER J K, SAUER F D, WOLYNETZ M S, et al. Effects of dietary saturated fat on erucic acid induced myocardial lipidosis in rats [J]. Lipids, 1992, 27 (8): 619-623.

[165] SAUER F D, KRAMER J K, FORESTER G V, et al. Palmitic and erucic acid metabolism in isolated perfused hearts from weanling pigs [J]. Biochimica et biophysica acta, 1989, 1004 (2): 205-214.

[166] YANG X, DICK T A. Dietary α-linolenic and linoleic acids competitively affect metabolism of polyunsaturated fatty acids in Arctic charr (*Salvelinus alpinus*) [J]. The Journal of nutrition, 1994, 124 (7): 1133-1145.

[167] ZENG Y Y, JIANG W D, LIU Y, et al. Dietary alpha-linolenic acid/linoleic acid ratios modulate intestinal immunity, tight junctions, anti-oxidant status and mRNA levels of NF-κB p65, MLCK and Nrf2 in juvenile grass carp (*Ctenopharyngodon idella*) [J]. Fish & shellfish immunology, 2016, 51: 351-364.

[168] WEN J, JIANG W, FENG L, et al. The influence of graded levels of available phosphorus on growth performance, muscle antioxidant and flesh quality of young grass carp (*Ctenopharyngodon idella*) [J]. Animal Nutrition, 2015, 1 (2): 77-84.

[169] 邵绪远. 饲粮纤维对生长中期草鱼生产性能、消化吸收能力和肠道健康的作用及其机制 [D]. 成都：四川农业大学，2019.

[170] JIANG W D, DENG Y P, LIU Y, et al. Dietary leucine regulates the intestinal immune status, immune-related signalling molecules and tight junction transcript abundance in grass carp (*Ctenopharyngodon idella*) [J]. Aquacul-

ture, 2015, 444: 134-142.

[171] STROBAND H W J, VD MEER H, TIMMERMANS L P M. Regional functional differentiation in the gut of the grasscarp, *Ctenopharyngodon idella* (Val.) [J]. Histochemistry, 1979, 64 (3): 235-249.

[172] CHE J, SU B, TANG B, et al. Apparent digestibility coefficients of animal and plant feed ingredients for juvenile *Pseudobagrus ussuriensis* [J]. Aquaculture nutrition, 2017, 23 (5): 1128-1135.

[173] AUSTRENG E. Digestibility deter mination in fish using chromic oxide marking and analysis of contents from different segments of the gastrointestinal tract [J]. Aquaculture, 1978, 13 (3): 265-272.

[174] XU J, WU P, JIANG W D, et al. Optimal dietary protein level improved growth, disease resistance, intestinal immune and physical barrier function of young grass carp (*Ctenopharyngodon idella*) [J]. Fish & shellfish immunology, 2016, 55: 64-87.

[175] ZHANG J X, GUO L Y, FENG L, et al. Soybean β-conglycinin induces inflammation and oxidation and causes dysfunction of intestinal digestion and absorption in fish [J]. PloS one, 2013, 8 (3).

[176] INTERNATIONAL A. (2005) Official methods of analysis of AOAC International, AOAC International.

[177] 甘雷. 不同盐度下尼罗罗非鱼幼鱼的脂肪营养生理研究 [D]. 上海：华东师范大学, 2016.

[178] FOLCH J, LEES M, STANLEY G H S. A simple method for the isolation and purification of total lipides from animal tissues [J]. Journal of biological chemistry, 1957, 226 (1): 497-509.

[179] LIU J, LIANG S, LIU X, et al. The absence of $ABCD_2$ sensitizes mice to disruptions in lipid metabolism by dietary erucic acid [J]. Journal of lipid research, 2012, 53 (6): 1071-1079.

[180] LI E, LIM C, KLESIUS P H, et al. Growth, body fatty acid composition, immune response, and resistance to Streptococcus iniae of hybrid tilapia, *Oreochromis niloticus × Oreochromis aureus*, fed diets containing various levels of linoleic and linolenic acids [J]. Journal of the World Aquaculture Society, 2013, 44 (1): 42-55.

[181] FALEIRO F, NARCISO L. Lipid dynamics during early development of *Hippocampus guttulatus* seahorses: searching for clues on fatty acid requirements [J]. Aquaculture, 2010, 307 (1-2): 56-64.

[182] KOKOU F, SARROPOULOU E, COTOU E, et al. Effects of graded dietary levels of soy protein concentrate supplemented with methionine and phosphate on the immune and antioxidant responses of gilthead sea bream (*Sparus aurata* L.) [J]. Fish & shellfish immunology, 2017, 64: 111-121.

[183] KUCUKBAY Z, YAZLAK H, SAHIN N, et al. Zinc picolinate supplementation decreases oxidative stress in rainbow trout (*Oncorhynchus mykiss*) [J]. Aquaculture, 2006, 257 (1-4): 465-469.

[184] CASTEX M, LEMAIRE P, WABETE N, et al. Effect of dietary probiotic Pediococcus acidilactici on antioxidant defences and oxidative stress status of shrimp *Litopenaeus stylirostris* [J]. Aquaculture, 2009, 294 (3-4): 306-313.

[185] LU X, WANG C, LIU B. The role of Cu/Zn-SOD and Mn-SOD in the immune response to oxidative stress and pathogen challenge in the clam *Meretrix meretrix* [J]. Fish & shellfish immunology, 2015, 42 (1): 58-65.

[186] PARRILLA TAYLOR D P, ZENTENO-SAV N T, MAGALL N-BARAJAS F J. Antioxidant enzyme activity in pacific whiteleg shrimp (*Litopenaeus vannamei*) in response to infection with white spot syndrome virus [J]. Aquaculture, 2013, 380: 41-46.

[187] TRENZADO C, HIDALGO M C, GARC A-GALLEGO M, et al. Antioxidant enzymes and lipid peroxidation in sturgeon Acipenser naccarii and trout *Oncorhynchus mykiss*. A comparative study [J]. Aquaculture, 2006, 254 (1-4): 758-767.

[188] HABIG W H, PABST M J, JAKOBY W B. Glutathione S-transferases the first enzymatic step in mercapturic acid formation [J]. Journal of biological chemistry, 1974, 249 (22): 7130-7139.

[189] CARLBERG I, MANNERVIK B. Purification and characterization of the flavoenzyme glutathione reductase from rat liver [J]. Journal of biological chemistry, 1975, 250 (14): 5475-5480.

[190] RITOLA O, LIVINGSTONE D R, PETERS L D, et al. Antioxidant processes are affected in juvenile rainbow trout (*Oncorhynchus mykiss*) exposed to ozone

and oxygen-supersaturated water [J]. Aquaculture, 2002, 210 (1-4): 1-19.

[191] KAWAKAMI M, INAGAWA R, HOSOKAWA T, et al. Mechanism of apoptosis induced by copper in PC12 cells [J]. Food and chemical toxicology, 2008, 46 (6): 2157-2164.

[192] SU J, ZHANG R, DONG J, et al. Evaluation of internal control genes for qRT-PCR normalization in tissues and cell culture for antiviral studies of grass carp (*Ctenopharyngodon idella*) [J]. Fish & shellfish immunology, 2011, 30 (3): 830-835.

[193] LIVAK K J, SCHMITTGEN T D. Analysis of relative gene expression data using real-time quantitative PCR and the $2^{-\Delta\Delta CT}$ method [J]. methods, 2001, 25 (4): 402-408.

[194] REN, X D. Regulation of the small GTP-binding protein Rho by cell adhesion and the cytoskeleton [J]. Embo Journal, 1999, 18 (3): 578-585.

[195] HOOFT J M, ELMOR A E H I, ENCARNA O P, et al. Rainbow trout (*Oncorhynchus mykiss*) is extremely sensitive to the feed-borne Fusarium mycotoxin deoxynivalenol (DON) [J]. Aquaculture, 2011, 311 (1-4): 224-232.

[196] ROBBINS K R, SAXTON A M, SOUTHERN L L. Estimation of nutrient requirements using broken-line regression analysis [J]. Journal of animal science, 2006, 84 (suppl_13): E155-E165.

[197] LI Y, YANG P, ZHANG Y, et al. Effects of dietary glycinin on the growth performance, digestion, intestinal morphology and bacterial community of juvenile turbot, *Scophthalmus maximus* L [J]. Aquaculture, 2017, 479: 125-133.

[198] LI Y, HU H, LIU J, et al. Dietary soya allergen β-conglycinin induces intestinal inflammatory reactions, serum-specific antibody response and growth reduction in a carnivorous fish species, turbot *Scophthalmus maximus* L [J]. Aquaculture research, 2017, 48 (8): 4022-4037.

[199] DENSTADLI V, SKREDE A, KROGDAHL Å, et al. Feed intake, growth, feed conversion, digestibility, enzyme activities and intestinal structure in Atlantic salmon (*Salmo salar* L.) fed graded levels of phytic acid [J]. Aquaculture, 2006, 256 (1-4): 365-376.

[200] KROGDAHL Å, GAJARDO K, KORTNER T M, et al. Soya saponins induce enteritis in Atlantic salmon (*Salmo salar* L.) [J]. Journal of agricultural and food chemistry, 2015, 63 (15): 3887-3902.

[201] CARROLL K K. Levels of radioactivity in tissues and in expired carbon dioxide after ad ministration of $1-C^{14}$-labelled palmitic acid, $2-C^{14}$-labelled erucic acid, or $2-C^{14}$-labelled nervonic acid to rats [J]. Canadian journal of biochemistry and physiology, 1962, 40 (9): 1229-1238.

[202] CHEN W, AI Q, MAI K, et al. Effects of dietary soybean saponins on feed intake, growth performance, digestibility and intestinal structure in juvenile Japanese flounder (*Paralichthys olivaceus*) [J]. Aquaculture, 2011, 318 (1-2): 95-100.

[203] LI L, LI M, ZHU R, et al. Effects of β-conglycinin on growth performance, antioxidant capacity and intestinal health in juvenile golden crucian carp, *Carassius auratus* [J]. Aquaculture research, 2019, 50 (11): 3231-3241.

[204] GU M, JIA Q, ZHANG Z, et al. Soya-saponins induce intestinal inflammation and barrier dysfunction in juvenile turbot (*Scophthalmus maximus*) [J]. Fish & shellfish immunology, 2018, 77: 264-272.

[205] WANG K Z, FENG L, JIANG W-D, et al. Dietary gossypol reduced intestinal immunity and aggravated inflammation in on-growing grass carp (*Ctenopharyngodon idella*) [J]. Fish & shellfish immunology, 2019, 86: 814-831.

[206] HANSEN P A, HAN D H, MARSHALL B A, et al. A high fat diet impairs stimulation of glucose transport in muscle functional evaluation of potential mechanisms [J]. Journal of biological chemistry, 1998, 273 (40): 26157-26163.

[207] SIT S P, NYHOF R, GALLAVAN R, et al. Mechanisms of glucose-induced hyperemia in the jejunum [J]. Proceedings of the Society for Experimental Biology and Medicine, 1980, 163 (2): 273-277.

[208] KAA E. In vitro biosynthesis of prostaglandin E 2 by kidney medulla of essential fatty acid deficient rats [J]. Lipids, 1976, 11 (9): 693-696.

[209] URIBE A, JOHANSSON C. Initial kinetic changes of prostaglandin E2-induced hyperplasia of the rat small intestinal epithelium occur in the villous compartments [J]. Gastroenterology, 1988, 94 (6): 1335-1342.

[210] KAMIYA S, NAGINO M, KANAZAWA H, et al. The value of bile replace-

ment during external biliary drainage: an analysis of intestinal permeability, integrity, and microflora [J]. Annals of surgery, 2004, 239 (4): 510.

[211] LUK G D, BAYLESS T M, BAYLIN S B. Plasma postheparin dia mine oxidase. Sensitive provocative test for quantitating length of acute intestinal mucosal injury in the rat [J]. The Journal of clinical investigation, 1983, 71 (5): 1308-1315.

[212] SMITH S M, ENG R H, CAMPOS J M, et al. D-lactic acid measurements in the diagnosis of bacterial infections [J]. Journal of clinical microbiology, 1989, 27 (3): 385-388.

[213] NIELSEN C, LINDHOLT J S, ERLANDSEN E J, et al. D-lactate as a marker of venous-induced intestinal ischemia: an experimental study in pigs [J]. International Journal of Surgery, 2011, 9 (5): 428-432.

[214] ZHANG Y L, JIANG W D, DUAN X D, et al. Soybean glycinin caused NADPH-oxidase-regulated ROS overproduction and decreased ROS eli mination capacity in the mid and distal intestine of juvenile grass carp (*Ctenopharyngodon idella*) [J]. Aquaculture, 2020, 516: 734651.

[215] DUAN X D, JIANG W D, WU P, et al. Soybean β-conglycinin caused intestinal inflammation and oxidative damage in association with NF-κB, TOR and Nrf2 in juvenile grass carp (*Ctenopharyngodon idella*): varying among different intestinal segments [J]. Fish & shellfish immunology, 2019, 95: 105-116.

[216] LI M, LI L, KONG Y D, et al. Effects of glycinin on growth performance, immunity and antioxidant capacity in juvenile golden crucian carp, *Cyprinus carpio* × *Carassius auratus* [J]. Aquaculture research, 2019, 51 (2): 465-479.

[217] LI L, LI M, ZHU R, et al. Effects of β-conglycinin on growth performance, antioxidant capacity and intestinal health in juvenile golden crucian carp, *Carassius auratus* [J]. Aquaculture research, 2019, 50 (11): 3231-3241.

[218] GIULIANI M E, REGOLI F. Identification of the Nrf2-Keap1 pathway in the European eel *Anguilla anguilla*: role for a transcriptional regulation of antioxidant genes in aquatic organisms [J]. Aquatic toxicology, 2014, 150: 117-123.

[219] CAI J, HUANG Y, WEI S, et al. Characterization of p38 MAPKs from orange-spotted grouper, *Epinephelus coioides* involved in SGIV infection [J]. Fish & shellfish immunology, 2011, 31 (6): 1129-1136.

[220] GARNER B, WALDECK A R, WITTING P K, et al. Oxidation of high density lipoproteins II. Evidence for direct reduction of lipid hydroperoxides by methionine residues of apolipoproteins AI and AII [J]. Journal of biological chemistry, 1998, 273 (11): 6088-6095.

[221] GANAPATHY E, SU F, MERIWETHER D, et al. D-4F, an apoA-I mimetic peptide, inhibits proliferation and tumorigenicity of epithelial ovarian cancer cells by upregulating the antioxidant enzyme MnSOD [J]. International journal of cancer, 2012, 130 (5): 1071-1081.

[222] NAWAZ M, MANZL C, LACHER V, et al. Copper-induced stimulation of extracellular signal-regulated kinase in trout hepatocytes: the role of reactive oxygen species, Ca^{2+}, and cell energetics and the impact of extracellular signal-regulated kinase signaling on apoptosis and necrosis [J]. Toxicological sciences: an official journal of the Society of Toxicology, 2006, 92 (2): 464-475.

[223] HRUBIK J, GLISIC B, FA S, et al. Erk-Creb pathway suppresses glutathione-S-transferase pi expression under basal and oxidative stress conditions in zebrafish embryos [J]. Toxicology Letters, 2016, 240 (1): 81-92.

[224] UCHIDA K, KANEMATSU M, MORIMITSU Y, et al. Acrolein is a product of lipid peroxidation reaction. Formation of free acrolein and its conjugate with lysine residues in oxidized low density lipoproteins [J]. The Journal of biological chemistry, 1998, 273 (26): 16058-16066.

[225] MCMAHON M, LAMONT D J, BEATTIE K A, et al. Keap1 perceives stress via three sensors for the endogenous signaling molecules nitric oxide, zinc, and alkenals [J]. Proceedings of the National Academy of Sciences of the United States of America, 2010, 107 (44): 18838-18843.

[226] LI L, KOBAYASHI M, KANEKO H, et al. Molecular evolution of Keap1-Two Keap1 molecules with distinctive intervening region structures are conserved among fish [J]. Journal of biological chemistry, 2008, 283 (6): 3248-3255.

[227] UMASUTHAN N, BATHIGE S, REVATHY K S, et al. A manganese superoxide dismutase (MnSOD) from *Ruditapes philippinarum*: comparative structural-and

expressional-analysis with copper/zinc superoxide dismutase (Cu/ZnSOD) and biochemical analysis of its antioxidant activities [J]. Fish & shellfish immunology, 2012, 33 (4): 753-765.

[228] LIU X, FENG J, XU Z R, et al. Oral allergy syndrome and anaphylactic reactions in BALB/c mice caused by soybean glycinin and β-conglycinin [J]. Clinical & Experimental Allergy, 2008, 38 (2): 350-356.

[229] MAT S J M, SEGURA J M, P REZ G MEZ C, et al. Antioxidant enzymatic activities in human blood cells after an allergic reaction to pollen or house dust mite [J]. Blood Cells, Molecules, and Diseases, 1999, 25 (2): 103-109.

[230] AKBARIAN A, MICHIELS J, DEGROOTE J, et al. Association between heat stress and oxidative stress in poultry: mitochondrial dysfunction and dietary interventions with phytochemicals [J]. Journal of Animal Science and Biotechnology, 2016, 7: 37.

[231] AZEKOSHI Y, YASU T, WATANABE S, et al. Free Fatty Acid Causes Leukocyte Activation and Resultant Endothelial Dysfunction Through Enhanced Angiotensin II Production in Mononuclear and Polymorphonuclear Cells [J]. Hypertension, 2010, 56 (1): 136-218.

[232] KIRIBAYASHI K, MASAKI T, NAITO T, et al. Angiotensin II induces fibronectin expression in human peritoneal mesothelial cells via ERK1/2 and p38 MAPK [J]. Kidney international, 2005, 67 (3): 1126-1135.

[233] JIANG L, WANG W, HE Q T, et al. Oleic acid induces apoptosis and autophagy in the treatment of Tongue Squamous cell carcinomas [J]. Scientific reports, 2017, 7: 11277.

[234] DENSTADLI V, VEGUSDAL A, KROGDAHL Å, et al. Lipid absorption in different segments of the gastrointestinal tract of Atlantic salmon (*Salmo salar* L.) [J]. Aquaculture, 2004, 240 (1-4): 385-398.

[235] RAHMAN H, NASREEN L, HABIB K, et al. Effects of Dietary Coconut Oil on Erucic Acid Rich Rapeseed Oil-induced Changes of Blood Serum Lipids in Rats [J]. Current Nutrition & Food Science, 2014, 10 (4): 302-307.

[236] CHOI B R, PALMQUIST D L. High fat diets increase plasma cholecystokinin and pancreatic polypeptide, and decrease plasma insulin and feed intake in lactating cows [J]. The Journal of nutrition, 1996, 126 (11): 2913-2919.

[237] GUKOVSKAYA A S, GUKOVSKY I, JUNG Y, et al. Cholecystokinin induces caspase activation and mitochondrial dysfunction in pancreatic acinar cells. Roles in cell injury processes of pancreatitis [J]. The Journal of biological chemistry, 2002, 277 (25): 22595-22604.

[238] FENG K, ZHANG G R, WEI K J, et al. Molecular characterization of cholecystokinin in grass carp (*Ctenopharyngodon idellus*): cloning, localization, developmental profile, and effect of fasting and refeeding on expression in the brain and intestine [J]. Fish physiology and biochemistry, 2012, 38 (6): 1825-1834.

[239] VASCONCELLOS F C S, WOICIECHOWSKI A L, SOCCOL V T, et al. Antimicrobial and antioxidant properties of-conglycinin and glycinin from soy protein isolate [J]. International Journal of Current Microbiology and Applied Sciences, 2014, 3 (8): 144-157.

[240] ZHAO Y, QIN G, SUN Z, et al. Disappearance of immunoreactive glycinin and [J]. Archives of animal nutrition, 2008, 62 (4): 322-330.

[241] GAN L, WU P, FENG L, et al. Erucic acid inhibits growth performance and disrupts intestinal structural integrity of on-growing grass carp (*Ctenopharyngodon idella*) [J]. Aquaculture, 2019, 513: 734437.

[242] MUSCH M W, WALSH REITZ M M, CHANG E B. Roles of ZO-1, occludin, and actin in oxidant-induced barrier disruption [J]. American Journal of Physiology Gastrointestinal & Liver Physiology, 2006, 290 (2): 222-231.

[243] TAMURA A, KITANO Y, HATA M, et al. Megaintestine in Claudin-15-Deficient Mice [J]. Gastroenterology, 2008, 134 (2): 523-534.

[244] SHEN L, BLACK E D, WITKOWSKI E D, et al. Myosin light chain phosphorylation regulates barrier function by remodeling tight junction structure [J]. Journal of cell science, 2006, 119 (10): 2095-2106.

[245] IVANOV A I, BACHAR M, BABBIN BRIAN A, et al. A Unique Role for Nonmuscle Myosin Heavy Chain IIA in Regulation of Epithelial Apical Junctions [J]. PloS one, 2007, 2 (8): e658.

[246] SAHAI E, MARSHALL C J. ROCK and Dia have opposing effects on adherens junctions downstream of Rho [J]. Nature cell biology, 2002, 4 (6): 408-415.

[247] TAMURA A, HAYASHI H, IMASATO M, et al. Loss of claudin-15, but not claudin-2, causes Na+ deficiency and glucose malabsorption in mouse small intestine [J]. Gastroenterology, 2011, 140 (3): 913-923.

[248] BOSSUS M C, MADSEN S S, TIPSMARK C K. Functional dynamics of claudin expression in Japanese medaka (Oryzias latipes): response to environmental salinity [J]. Comparative Biochemistry and Physiology Part A: Molecular & Integrative Physiology, 2015, 187: 74-85.

[249] FUJITA H, SUGIMOTO K, INATOMI S, et al. Tight junction proteins claudin-2 and -12 are critical for vita min D-dependent Ca^{2+} absorption between enterocytes [J]. Molecular biology of the cell, 2008, 19 (5): 1912-1921.

[250] LI T, KOSHY S, FOLKESSON H G. IL-1β-induced cortisol stimulates lung fluid absorption in fetal guinea pigs via SGK-mediated Nedd4-2 inhibition [J]. American Journal of Physiology-Lung Cellular and Molecular Physiology, 2009, 296 (3): L527-L533.

[251] BUI P, BAGHERIE LACHIDAN M, KELLY S P. Cortisol differentially alters claudin isoforms in cultured puffer fish gill epithelia [J]. Molecular and cellular endocrinology, 2010, 317 (1-2): 120-126.

[252] LANFRANCO F, GIORDANO R, PELLEGRINO M, et al. Free fatty acids exert an inhibitory effect on adrenocorticotropin and cortisol secretion in humans [J]. The Journal of Clinical Endocrinology & Metabolism, 2004, 89 (3): 1385-1390.

[253] BRUEWER M, LUEGERING A, KUCHARZIK T, et al. Proinflammatory cytokines disrupt epithelial barrier function by apoptosis-independent mechanisms [J]. The Journal of Immunology, 2003, 171 (11): 6164-6172.

[254] CHANG Y W E, MARLIN J W, CHANCE T W, et al. RhoA mediates cyclooxygenase-2 signaling to disrupt the formation of adherens junctions and increase cell motility [J]. Cancer research, 2006, 66 (24): 11700-11708.

[255] WANG X M, ZHAO X H, FENG T, et al. Rutin Prevents High Glucose-Induced Renal Glomerular Endothelial Hyperpermeability by Inhibiting the ROS/Rhoa/ROCK Signaling Pathway [J]. Planta Medica, 2016, 82 (14): 1252-1257.

[256] MULIER B, RAHMAN I, WATCHORN T, et al. Hydrogen peroxide-induced

epithelial injury: the protective role of intracellular nonprotein thiols (NPSH) [J]. European Respiratory Journal, 1998, 11 (2): 384-391.

[257] 冯琳. 大豆凝集素对鲤鱼肠道上皮细胞增殖分化及其功能的影响 [D]. 雅安: 四川农业大学, 2006.

[258] JIANG W D, LIU Y, HU K, et al. Copper exposure induces oxidative injury, disturbs the antioxidant system and changes the Nrf2/ARE (CuZnSOD) signaling in the fish brain: protective effects of myo-inositol [J]. Aquatic toxicology, 2014, 155: 301-313.

[259] LI C, XU H, XIAO L, et al. CRMP4a suppresses cell motility by sequestering RhoA activity in prostate cancer cells [J]. Cancer Biology & Therapy, 2018, 19 (12): 1193-1203.

[260] FUKUTA T, NISHIKAWA A, KOGURE K. Low level electricity increases the secretion of extracellular vesicles from cultured cells [J]. Biochemistry and Biophysics Reports, 2020, 21: 100713.

[261] LIU P, ZHU C, LUO J, et al. Par6 regulates cell cycle progression through enhancement of Akt/PI3K/GSK-3β signaling pathway activation in glioma [J]. The FASEB Journal, 2020, 34 (1): 1481-1496.

[262] LI C, XU H, XIAO L, et al. CRMP4a suppresses cell motility by sequestering RhoA activity in prostate cancer cells [J]. Cancer Biology & Therapy, 2018, 19 (12): 1193-1203.

[263] PENG J H, LENG J, TIAN H J, et al. Geniposide and Chlorogenic Acid Combination Ameliorates Non-alcoholic Steatohepatitis Involving the Protection on the Gut Barrier Function in Mouse Induced by High-Fat Diet [J]. Frontiers in pharmacology, 2018, 9: 1399.

[264] LE DREAN G, HAURE MIRANDE V, FERRIER L, et al. Visceral adipose tissue and leptin increase colonic epithelial tight junction permeability via a RhoA-ROCK-dependent pathway [J]. Faseb Journal, 2014, 28 (3): 1059-1070.

[265] CRESPO I, SAN MIGUEL B, PRAUSE C, et al. Gluta mine treatment attenuates endoplasmic reticulum stress and apoptosis in TNBS-induced colitis [J]. PloS one, 2012, 7 (11): e50407.

[266] BEARE ROGERS J L, NERA E A, CRAIG B M. Accumulation of cardiac fatty

acids in rats fed synthesized oils containing C_{22} fatty acids [J]. Lipids, 1972, 7 (1): 46-50.

[267] CUI W, MA J, WANG X, et al. Free fatty acid induces endoplasmic reticulum stress and apoptosis of β-cells by Ca^{2+}/calpain-2 pathways [J]. PloS one, 2013, 8 (3): e59921.

[268] SONG Y F, LUO Z, CHEN Q L, et al. Protective Effects of Calcium Pre-Exposure Against Waterborne Cadmium Toxicity in Synechogobius hasta [J]. Archives of Environmental Conta mination and Toxicology, 2013, 65 (1): 105-121.

[269] WANG Y, WU T, TANG M. Ambient particulate matter triggers dysfunction of subcellular structures and endothelial cell apoptosis through disruption of redox equilibrium and calcium homeostasis [J]. Journal of hazardous materials, 2020, 394: 122439.

[270] TRAN H, MITTAL A, SAGI V, et al. Mast Cells Induce Blood Brain Barrier Damage in SCD by Causing Endoplasmic Reticulum Stress in the Endothelium [J]. Frontiers in Cellular Neuroscience, 2019, 13: 56.

[271] SHAO B, WANG M, CHEN A, et al. Protective effect of caffeic acid phenethyl ester against imidacloprid-induced hepatotoxicity by attenuating oxidative stress, endoplasmic reticulum stress, inflammation and apoptosis [J]. Pesticide biochemistry and physiology, 2020, 164: 122-129.

[272] YANG X, SHAO H L, LIU W R, et al. Endoplasmic reticulum stress and oxidative stress are involved in ZnO nanoparticle-induced hepatotoxicity [J]. Toxicology Letters, 2015, 234 (1): 40-49.

[273] MENG X, DONG H H, PAN Y W, et al. Diosgenyl Saponin Inducing Endoplasmic Reticulum Stress and Mitochondria-Mediated Apoptotic Pathways in Liver Cancer Cells [J]. Journal of agricultural and food chemistry, 2019, 67 (41): 11428-11435.

[274] AKIYAMA T, OISHI K, WULLAERT A. Bifidobacteria prevent tunicamycin-induced endoplasmic reticulum stress and subsequent barrier disruption in human intestinal epithelial Caco-2 monolayers [J]. PloS one, 2016, 11 (9): e0162448.

[275] LIN R, SUN Y, MU P, et al. Lactobacillus rhamnosus GG supplementation

modulates the gut microbiota to promote butyrate production, protecting against deoxynivalenol exposure in nude mice [J]. Biochemical pharmacology, 2020, 175: 113868.

[276] GULHANE M, MURRAY L, LOURIE R, et al. High fat diets induce colonic epithelial cell stress and inflammation that is reversed by IL-22 [J]. European Journal of Immunology, 2016, 46: 355-356.

[277] LIN R Q, SUN Y, YE W C, et al. T-2 toxin inhibits the production of mucin via activating the IRE1/XBP1 pathway [J]. Toxicology, 2019, 424: 152230.

[278] MAINGAT F, HALLORAN B, ACHARJEE S, et al. Inflammation and epithelial cell injury in AIDS enteropathy: involvement of endoplasmic reticulum stress [J]. Faseb Journal, 2011, 25 (7): 2211-2220.

[279] NIVALA A M, REESE L, FRYE M, et al. Fatty acid-mediated endoplasmic reticulum stress in vivo: differential response to the infusion of Soybean and Lard Oil in rats [J]. Metabolism, 2013, 62 (5): 753-760.

[280] WEI Y, WANG D, TOPCZEWSKI F, et al. Saturated fatty acids induce endoplasmic reticulum stress and apoptosis independently of ceramide in liver cells [J]. American Journal of Physiology Endocrinology & Metabolism, 2006, 291 (2): E275-E281.

[281] DIAKOGIANNAKI E, WELTERS H J, MORGAN N G. Differential regulation of the endoplasmic reticulum stress response in pancreatic β-cells exposed to long-chain saturated and monounsaturated fatty acids [J]. Journal of Endocrinology, 2008, 197 (3): 553-563.

[282] CAO J, DAI D L, YAO L, et al. Saturated fatty acid induction of endoplasmic reticulum stress and apoptosis in human liver cells via the PERK/ATF4/CHOP signaling pathway [J]. Molecular & Cellular Biochemistry, 2012, 364 (1-2): 115-129.

[283] D SIR VIGN A, HAURE MIRANDE V, DE COPPET P, et al. Perinatal supplementation of 4-Phenylbutyrate and gluta mine attenuates endoplasmic reticulum stress and improves colonic epithelial barrier function in rats born with intrauterine growth restriction [J]. Journal of Nutritional Biochemistry, 2018, 55: 104-112.

[284] XUE A, LIN J, QUE C, et al. Aberrant endoplasmic reticulum stress mediates

coronary artery spasm through regulating MLCK/MLC2 pathway [J]. Experimental Cell Research, 2018, 363 (2): 321-331.

[285] SONG Y F, HUANG C, SHI X, et al. Endoplasmic reticulum stress and dysregulation of calcium homeostasis mediate Cu-induced alteration in hepatic lipid metabolism of javelin goby Synechogobius hasta [J]. Aquatic toxicology, 2016, 175: 20-29.

[286] DATTA D, KHATRI P, SINGH A, et al. Mycobacterium fortuitum-induced ER-Mitochondrial calcium dynamics promotes calpain/caspase-12/caspase-9 mediated apoptosis in fish macrophages [J]. Cell death discovery, 2018, 4 (1): 1-13.

[287] THAKUR P C, DAVISON J M, STUCKENHOLZ C, et al. Dysregulated phosphatidylinositol signaling promotes endoplasmic-reticulum-stress-mediated intestinal mucosal injury and inflammation in zebrafish [J]. Disease models & mechanisms, 2014, 7 (1): 93-106.

[288] CHEN K, LI X, SONG G, et al. Deficiency in the membrane protein Tmbim3a/Grinaa initiates cold-induced ER stress and cell death by activating an intrinsic apoptotic pathway in zebrafish [J]. Journal of biological chemistry, 2019, 294 (30): 11445-11457.

[289] LING S C, WU K, ZHANG D G, et al. Endoplasmic Reticulum Stress-Mediated Autophagy and Apoptosis Alleviate Dietary Fat-Induced Triglyceride Accumulation in the Intestine and in Isolated Intestinal Epithelial Cells of Yellow Catfish [J]. The Journal of nutrition, 2019, 149 (10): 1732-1741.

[290] JIANG S, FAN Q, XU M, et al. Hydrogen-rich saline protects intestinal epithelial tight junction barrier in rats with intestinal ischemia-reperfusion injury by inhibiting endoplasmic reticulum stress-induced apoptosis pathway [J]. Journal of Pediatric Surgery, 2020 (in press).

[291] ZHOU Y, YE L, ZHENG B, et al. Phenylbutyrate prevents disruption of blood-spinal cord barrier by inhibiting endoplasmic reticulum stress after spinal cord injury [J]. American journal of translational research, 2016, 8 (4): 1864-1875.

[292] VAN DER GIESSEN J, VAN DER WOUDE C J, PEPPELENBOSCH M P, et al. A Direct Effect of Sex Hormones on Epithelial Barrier Function in Inflamma-

tory Bowel Disease Models [J]. Cells, 2019, 8 (3): 261.

[293] CHOTIKATUM S, NAIM H Y, EL NAJJAR N. Inflammation induced ER stress affects absorptive intestinal epithelial cells function and integrity [J]. International Immunopharmacology, 2018, 55: 336-344.

[294] 严锋. 内质网应激 PERK 通路在蛛网膜下腔出血早期脑损伤中的作用及机制研究 [D]. 杭州: 浙江大学, 2017.

[295] 郑晨果. IRE1α-XBP1 通路引起结直肠癌细胞增生、侵袭、转移的相关机制研究 [D]. 武汉: 武汉大学, 2017.

[296] 杨超. 高血糖激活 Kupffer 细胞 ATF6-CHOP 通路在肝脏缺血再灌注损伤中的作用及其机制研究 [D]. 南京: 南京医科大学, 2018.

[297] TANG X, LIANG X J, LI M H, et al. ATF6 pathway of unfolded protein response mediates advanced oxidation protein product-induced hypertrophy and epithelial-to-mesenchymal transition in HK-2 cells [J]. Molecular and cellular biochemistry, 2015, 407 (1-2): 197-207.

[298] PASINI S, LIU J, CORONA C, et al. Activating Transcription Factor 4 (ATF4) modulates Rho GTPase levels and function via regulation of RhoGDIα [J]. Scientific reports, 2016, 6: 36952.

[299] DING Q, ZHANG Z, RAN C, et al. The hepatotoxicity of palmitic acid in zebrafish involves the intestinal microbiota [J]. The Journal of nutrition, 2018, 148 (8): 1217-1228.

[300] LIU W, WEN Y, WANG M, et al. Enhanced Resistance of Triploid Crucian Carp to Cadmiu minduced Oxidative and Endoplasmic Reticulum Stresses [J]. Current molecular medicine, 2018, 18 (6): 400-408.

[301] SONG Y F, HOGSTRAND C, WEI C C, et al. Endoplasmic reticulum (ER) stress and cAMP/PKA pathway mediated Zn-induced hepatic lipolysis [J]. Environmental Pollution, 2017, 228: 256-264.

[302] AXTEN J M, ROMERIL S P, SHU A, et al. Discovery of GSK2656157: an optimized PERK inhibitor selected for preclinical development [J]. ACS medicinal chemistry letters, 2013, 4 (10): 964-968.

[303] ANDO T, KOMATSU T, NAIKI Y, et al. GSK2656157, a PERK inhibitor, reduced LPS-induced IL-1β production through inhibiting Caspase 1 activation in macrophage-like J774. 1 cells [J]. Immunopharmacology and immunotoxicol-

ogy, 2016, 38 (4): 298-302.

[304] SUN J, YU X, HUANGPU H, et al. Ginsenoside Rb3 protects cardiomyocytes against hypoxia/reoxygenation injury via activating the antioxidation signaling pathway of PERK/Nrf2/HMOX1 [J]. Biomedicine & Pharmacotherapy, 2019, 109: 254-261.

[305] MA B, ZHANG J, ZHU Z, et al. Aucubin, a natural iridoid glucoside, attenuates oxidative stress-induced testis injury by inhibiting JNK and CHOP activation via Nrf2 up-regulation [J]. Phytomedicine, 2019, 64: 153057.

[306] LIN Y, ZHANG C, XIANG P P, et al. Exosomes derived from HeLa cells break down vascular integrity by triggering endoplasmic reticulum stress in endothelial cells [J]. Journal of Extracellular Vesicles, 2020, 9 (1): 1722385.

[307] FAN L F, HE P Y, PENG Y C, et al. Mdivi-1 ameliorates early brain injury after subarachnoid hemorrhage via the suppression of inflammation-related blood-brain barrier disruption and endoplasmic reticulum stress-based apoptosis [J]. Free Radical Biology and Medicine, 2017, 112: 336-349.

[308] SIDRAUSKI C, WALTER P. The transmembrane kinase Ire1p is a site-specific endonuclease that initiates mRNA splicing in the unfolded protein response [J]. Cell, 1997, 90 (6): 1031-1039.

[309] COMINACINI L, MOZZINI C, GARBIN U, et al. Endoplasmic reticulum stress and Nrf2 signaling in cardiovascular diseases [J]. Free Radical Biology and Medicine, 2015, 88: 233-242.

[310] ABUAITA B H, BURKHOLDER K M, BOLES B R, et al. The Endoplasmic Reticulum Stress Sensor Inositol-Requiring Enzyme 1α Augments Bacterial Killing through Sustained Oxidant Production [J]. Mbio, 2015, 6 (4): e00705-00715.

[311] FAN X, LI S, WU Z, et al. Glycine supplementation to breast-fed piglets attenuates post-weaning jejunal epithelial apoptosis: a functional role of CHOP signaling [J]. Amino acids, 2019, 51 (3): 463-473.

[312] WU K, LUO Z, HOGSTRAND C, et al. Zn Stimulates the Phospholipids Biosynthesis via the Pathways of Oxidative and Endoplasmic Reticulum Stress in the Intestine of Freshwater Teleost Yellow Catfish [J]. Environmental Science & Technology, 2018, 52 (16): 9206-9214.

[313] LIU L, XU L, ZHANG S, et al. STF-083010, an inhibitor of XBP1 splicing, attenuates acute renal failure in rats by suppressing endoplasmic reticulum stress-induced apoptosis and inflammation [J]. Experimental animals, 2018: 17-0131.

[314] CHEN Q Q, ZHANG C, QIN M Q, et al. Inositol-requiring enzyme 1 alpha endoribonuclease specific inhibitor STF-083010 alleviates carbon tetrachloride induced liver injury and liver fibrosis in mice [J]. Frontiers in pharmacology, 2018, 9: 1344.

[315] ZHAN F, ZHAO G, LI X, et al. Inositol-requiring enzyme 1 alpha endoribonuclease specific inhibitor STF-083010 protects the liver from thioacetamide-induced oxidative stress, inflammation and injury by triggering hepatocyte autophagy [J]. International Immunopharmacology, 2019, 73: 261-269.

[316] CUEVAS E P, ERASO P, MAZ N M J, et al. LOXL2 drives epithelial-mesenchymal transition via activation of IRE1-XBP1 signalling pathway [J]. Scientific reports, 2017, 7: 44988.

[317] YOSHIDA H, HAZE K, YANAGI H, et al. Identification of the cis-acting endoplasmic reticulum stress response element responsible for transcriptional induction of mammalian glucose-regulated proteins Involvement of basic leucine zipper transcription factors [J]. Journal of biological chemistry, 1998, 273 (50): 33741-33749.

[318] KOKAME K, KATO H, MIYATA T. Identification of ERSE-II, a new cis-acting element responsible for the ATF6-dependent mammalian unfolded protein response [J]. Journal of biological chemistry, 2001, 276 (12): 9199-9205.

[319] TOKO H, TAKAHASHI H, KAYAMA Y, et al. ATF6 is important under both pathological and physiological states in the heart [J]. Journal of molecular and cellular cardiology, 2010, 49 (1): 113-120.

[320] TAM A B, ROBERTS L S, CHANDRA V, et al. The UPR activator ATF6 responds to proteotoxic and lipotoxic stress by distinct mechanisms [J]. Developmental cell, 2018, 46 (3): 327-343.

[321] LEBEAU P, BYUN J H, YOUSOF T, et al. Pharmacologic inhibition of S1P attenuates ATF6 expression, causes ER stress and contributes to apoptotic cell death [J]. Toxicology and applied pharmacology, 2018, 349: 1-7.

[322] CHEN J, ZHANG M, ZHU M, et al. Paeoniflorin prevents endoplasmic reticulum stress-associated inflammation in lipopolysaccharide-stimulated human umbilical vein endothelial cells via the IRE1α/NF-κB signaling pathway [J]. Food & function, 2018, 9 (4): 2386-2397.

[323] HOSOI T, HYODA K, OKUMA Y, et al. Geldanamycin induces CHOP expression through a 4-(2-a minoethyl) -benzenesulfonyl fluoride-responsive serine protease [J]. Cell research, 2007, 17 (2): 184-186.

[324] DADEY D Y A, KAPOOR V, KHUDANYAN A, et al. The ATF6 pathway of the ER stress response contributes to enhanced viability in glioblastoma [J]. Oncotarget, 2016, 7 (2): 2080.

[325] TAY K H, LUAN Q, CROFT A, et al. Sustained IRE1 and ATF6 signaling is important for survival of melanoma cells undergoing ER stress [J]. Cellular signalling, 2014, 26 (2): 287-294.

[326] YOSHIKAWA T, OGATA N, IZUTA H, et al. Increased expression of tight junctions in ARPE-19 cells under endoplasmic reticulum stress [J]. Current eye research, 2011, 36 (12): 1153-1163.

[327] LENIN R, NAGY P G, JHA K A, et al. GRP78 translocation to the cell surface and O-GlcNAcylation of VE-Cadherin contribute to ER stress-mediated endothelial permeability [J]. Scientific reports, 2019, 9 (1): 10783.